33TH ANNIVERSARY OF PEARL RIVER DESIGN (VOLUME 1)

珠江设计33 上

广州珠江外资建筑设计院有限公司　主编

CHIEF EDITOR: GUANGZHOU PEARL RIVER FOREIGN INVESTMENT ARCHITECTURAL DESIGNING INSTITUTE CO. LTD

中国建筑工业出版社
CHINA ARCHITECTURE & BUILDING PRESS

前言　面向未来的岭南新建筑创作

历史上，岭南人在岭南建筑文化上勇于探索、敢为人先，从而使岭南传统建筑不断有创新的形态出现，只要“万物皆备于我”，便可接纳融合。岭南建筑从适应社会生活的实际出发，按不同的时空条件而变化，不拘泥于传统形制和模式，尽量采用新技术、新材料、新设备和新颖的形式，以体现时代气息。因此岭南建筑文化总能开风气之先，不断创新。

设计展示其独特文化底蕴的岭南新建筑，离不开岭南地域文化原创性这一精神原点。对地域文化的提炼、抽象和创新，是岭南建筑文化延续和发展的精髓。基于地域文化的岭南新建筑创作，就是要既立足于岭南，做到融会贯通、形神兼备，回归岭南生活世界，又不拘泥于传统形式，勇于创新变革，以超越模拟的象征手法，运用现代高科技手段表达出地域文化的创新之美。在其中多文化融合共生，给予人们视觉冲击力和无限想象的空间，将文化艺术的感染力引入审美。引发人们在岭南建筑文化背景下的共鸣，引导他们参与其中，获得其独特的地域文化体验。

基于自然环境的岭南新建筑创作就是将建筑师对自然的理解和情感用建筑语言表现出来，这实际包含一种回归自然、体现人与自然的和谐共存的艺术情结。将自然意象融入建筑的灵魂，在较深层次上探索和表现建筑与自然的关系。

岭南新建筑的自然意象的表达，是与信息时代的生态、可持续发展思想的深入人心分不开的。基于科技的岭南建筑文化创新就是通过高科技手段以及新材料、节能环保技术的运用，使建筑既散发出科技的魅力，又蕴涵着人类情感的活力，激发人们的好奇心与探索欲，日新月异的建筑技术和材料在岭南新建筑上得以最充分的展示。

探索岭南建筑文化特质的再现与重构，力求以传统的色彩、地方的材料，经过抽象的细部，传统建筑元素的现代表达等手法去体现岭南建筑文化的精神内涵。以现代的建筑语言创造符合现代岭南人生活方式的建筑空间，使人居环境既有现代时尚感又有历史价值感，赋予当代岭南新建筑以鲜明的地域特色。

面向未来的岭南新建筑创作就是要密切人与城市公共空间的相互关联，强化市民对城市文化的归属感和认同感，致力于创造崭新的建筑语言，不应只重复历史形式，应力求为城市和建筑开创一种新的语境。对于历史文脉的延续和发展，不追求简单的形式符号的借用，而是发掘并强调城市无形的、内在的特征与脉络，对城市建筑文脉进行“提炼”，贯穿岭南城市的历史和未来，赋予其抽象而全新的形态和情调，创造出富于时代感和生命力的岭南新建筑。

岭南建筑设计创新首先是观念和文化的创新，其次是手法和技术的创新。在适应自然、融合共生的基础上探索建筑与地域文化的关系，传承与创新岭南建筑文化是我们在建筑创作上不懈的追求。

珠江设计33　编辑组

目 录

方案创作类 SCHEME DESIGN

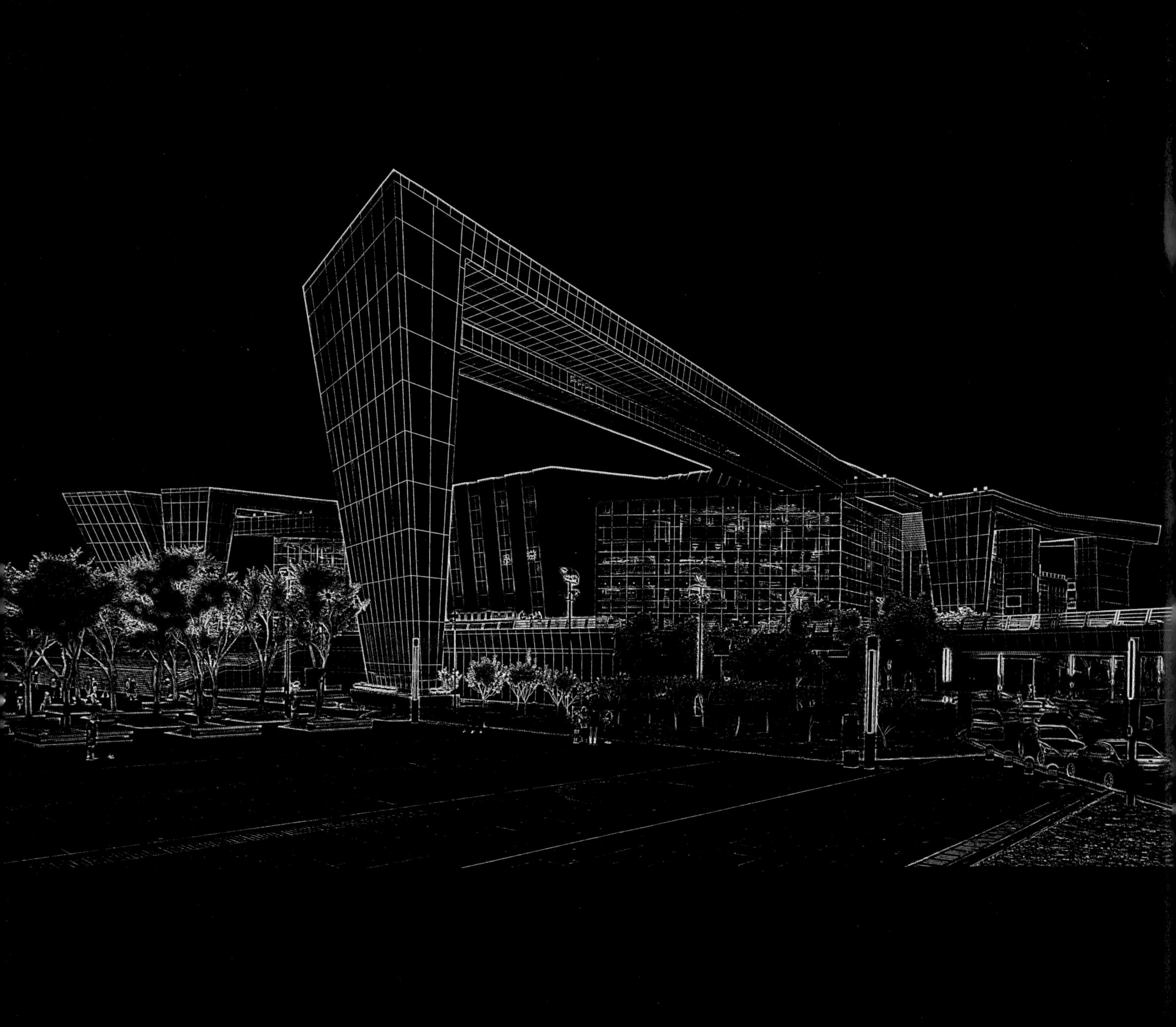

PUBLIC AND CULTURAL BUILDINGS
公共与文化类

01 武汉琴台文化艺术中心
Wuhan Qintai Culture and Art Center

琴台文化艺术中心占地15万平方米，总建筑面积约10.5万平方米，由琴台大剧院和琴台音乐厅组成，因与环境共生而成其大气，与空间相连而生其变幻，与文化相映而立其意境，形意相随，如高山放歌、流水知音，其音旷达，其境高远。

琴台大剧院以琴键飞奔、水袖飞舞般的飘带伸臂构架造型，限定出层次丰富的景观空间，与宛如古琴琴弦的金属玻璃体交相辉映，观众在走近大剧院的过程中，建筑给人的感受时如水袖飞舞，舒展飘逸，时如金石顿开，激情迸发。立面选材中，大胆地将楚风浮雕与质朴无华的清水混凝土、锈蚀的青铜和透明的玻璃并置碰撞，外部界面富于力度之美，内部空间如群山叠嶂，峰回路转，营造出高山流水般的激昂、奔

Planned with a site area of 150,000 sqm and a gross floor area of 105,000 sqm, Wuhan Qintai Culture & Art Center consists of Qintai Opera House and Qintai Concert. Its symbiotic relationship with the surroundings defines its majestic spirit. Its connections in space produce changes. Its combination with local culture achieves its finest sentimental effects. Form and theme go together in harmony here, which may inspire in you the picture of singing in mountains and seeing flowing streams, and bring you the broad-mindedness and unconventional tranquility.

Qintai Opera House is built with a cantilever framework that resembles instrument keys being played and dancing sleeves, defining a landscape space with abundant layers, which contrasts pleasantly with the metallic glass body that resembles Guqin (a representative instrument of traditional Chinese musical culture) strings, graceful and unrestraint in style, suggestive of a burst of passion. The facade is bold in style with the combined use of Chu-style (ancient Hubei features) relief, unaffected as-cast finish concrete, corroded bronze and transparent glass. The externals of the House presents a visual beauty of strength, while its interior space

放的楚风意境。

琴台音乐厅整体造型追求“行云流水、轻歌蔓舞”般的音乐感，强调线条在建筑中的巧妙组合。通过对清水混凝土挂板、铜板、玻璃等几种不同材料的使用形成材质及视觉上的对比，表达了“知音契合，岁月留痕”的意境。屋面蜿蜒曲折，中部有八片连续曲面金属板、七条玻璃天窗穿插其中，恰似流水，仿如琴弦，入口处十六片立柱构架，有编磬之意。

奖项：

2007 年度广东省注册建筑师协会优秀建筑创作奖；

2008 年度广州市优秀工程设计一等奖；

2008 年度武汉市科技进步一等奖；

is rugged, towering and winding, creating a passionate and unrestrained style typical of ancient Hubei, which may evoke the image of rushing mountain torrents. Qintai Concert is built in a smooth and graceful style that resembles sailing clouds and flowing water by emphasizing the ingenious combination of lines in architectural design. The use of as-cast finish concrete formwork, copper plate, glass and other materials creates a visual contrast, expressing the idea of natural fitness and age-old bearing. The roof features twists and turns, with eight consecutive curved metal plates in the middle part and seven glass sunroofs interspersed in it, which is suggestive of the image of flowing water and music instrument strings. At the entrance, 16 pillars are built, which resembles the stone and jade chimes, an ancient Chinese percussion instrument that is played in a unique way.

Awards:

Architectural Innovation Prize awarded by Guangdong Chapter of Association of Chinese Registered Architects in 2007.

1st Prize for Excellent Design awarded by Guangzhou Municipal Government in 2008.

1st Prize for Technology Progress awarded by Wuhan Municipal Government in 2008.

2009 年度湖北省科技进步一等奖；
2009 年度广东省优秀工程勘察设计一等奖；
2009 年度建设部全国优秀工程设计二等奖；
2009 年度广东省优秀工程设计一等奖；
2010 年度广州市优秀工程勘察设计大奖；
2010 年度广州市优秀工程勘察设计一等奖；
2011 年度广东省优秀工程勘察设计一等奖；
第十届中国土木工程詹天佑奖。

1st Prize for Technology Progress awarded by Hubei Provincial Government in 2009.
1st Prize for Construction Survey Design awarded by Guangdong Provincial Government in 2009.
2nd Prize for National Excellent Design awarded by the Ministry of Construction in 2009.
1st Prize for Excellent Design awarded by Guangdong Provincial Government in 2009.
Prize for Construction Survey Design awarded by Guangzhou Municipal Government in 2009.
1st Prize for Construction Survey Design awarded by Guangzhou Municipal Government in 2010.
Prize for Construction Survey Design awarded by Guangdong Provincial Government in 2011.
The 10th Tien-yow Jeme Civil Engineering Prize.

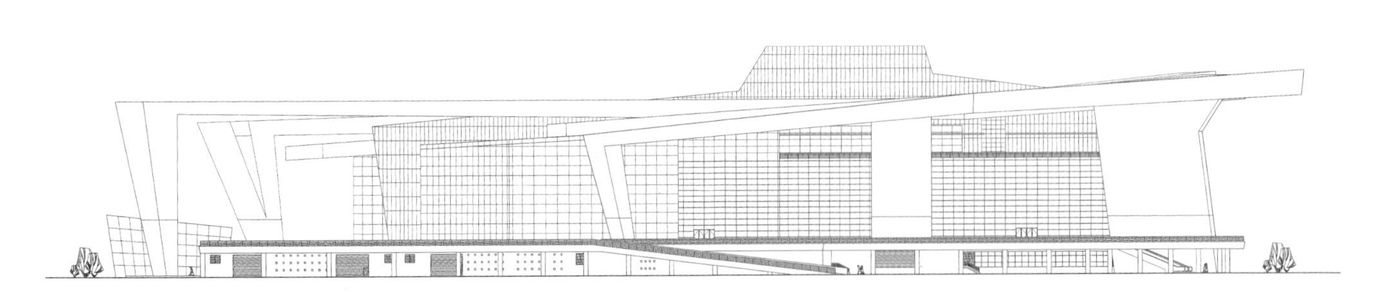

北立面图 Illustration of the North Facade

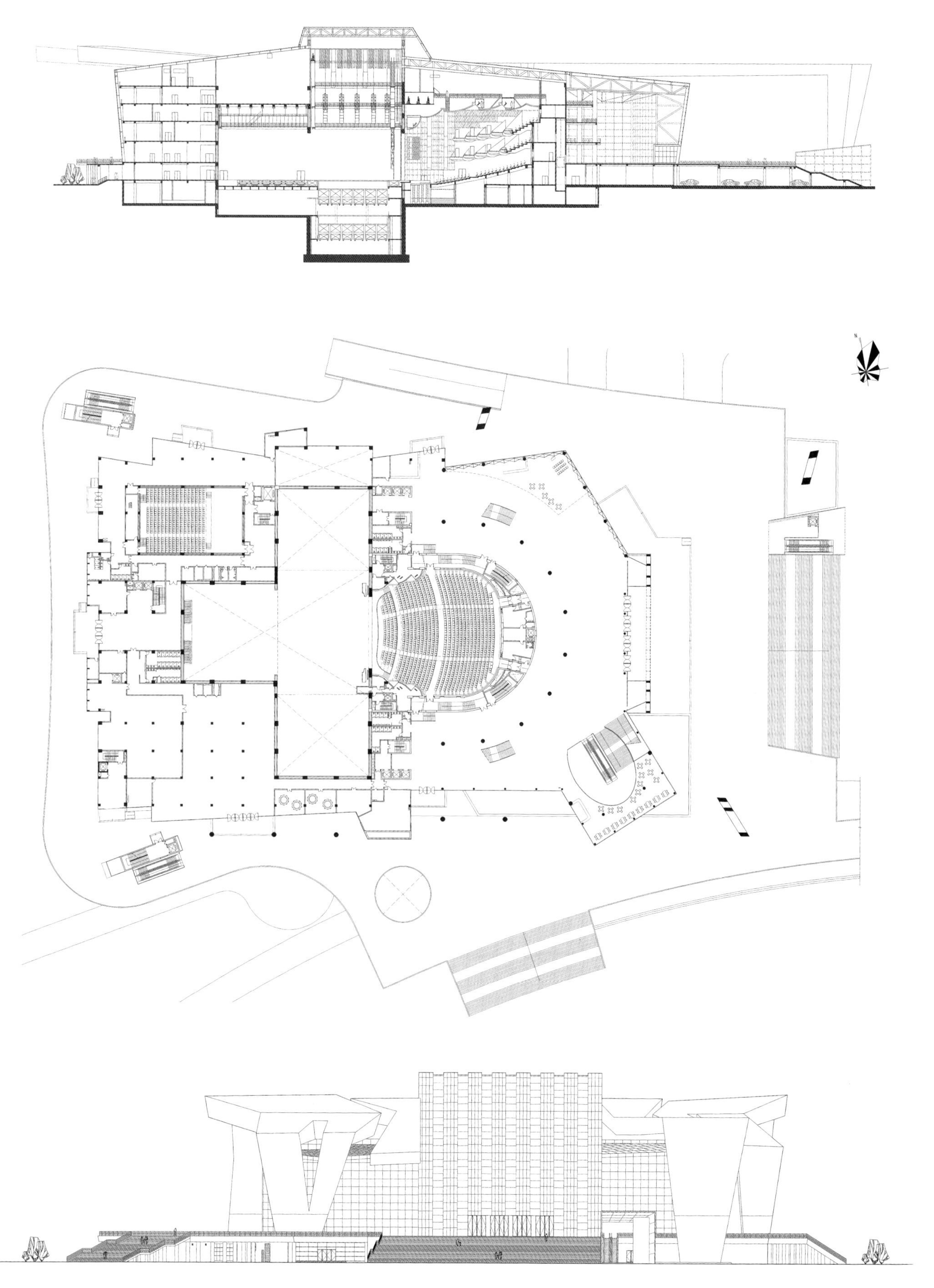

上至下：观众厅剖面图、5.4m标高平面图、东立面图。
From top to bottom, Section of the Auditorium, Plan of 5.4m Elevation Floor, Illustration of the East Facade.

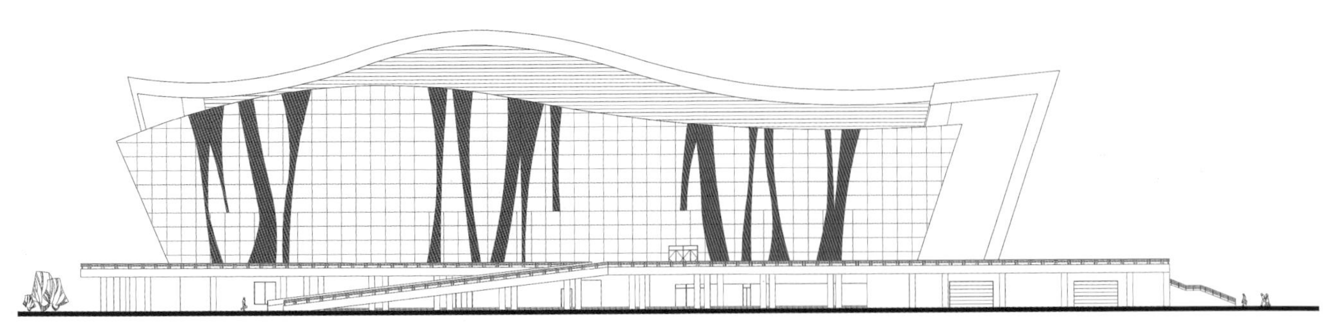

北立面图 Illustration of the North Facade

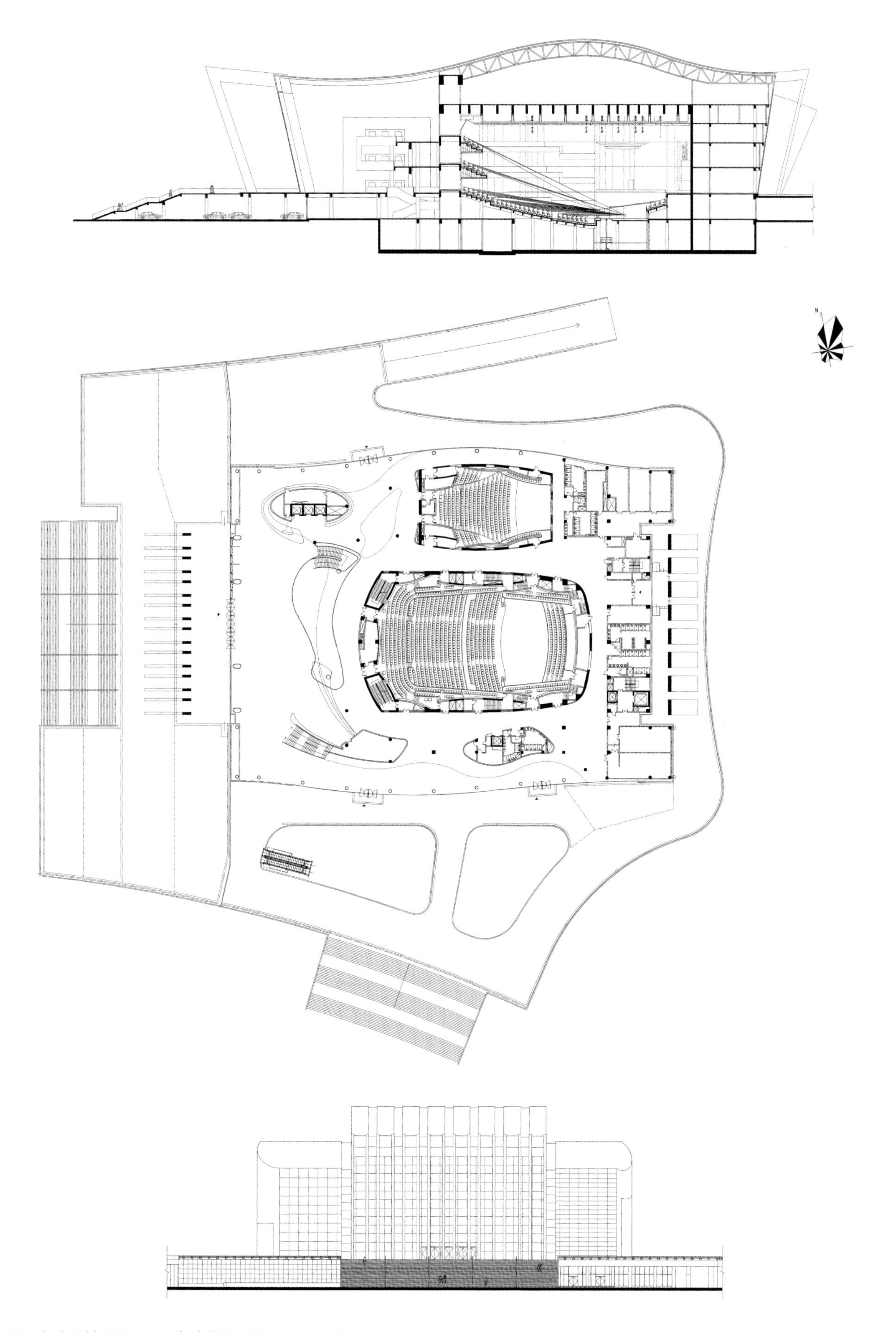

上至下：音乐厅剖面图、5.4m标高层平面图、西立面图。
From top to bottom, Section of the Concert, Plan of 5.4m Elevation, Illustration of the West Facade.

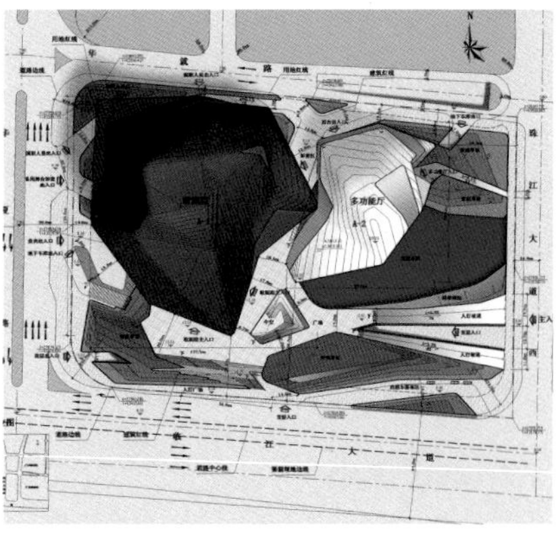

02 广州歌剧院
（与扎哈·哈迪德建筑师事务所合作设计）
Guangzhou Opera House
(In collaboration with Zaha Hadid Architects)

广州歌剧院位于新城市中轴线与珠江北岸交汇处的西侧，中轴线的东侧是广州图书馆和广东省博物馆，用地的北面是广州市第二少年宫，用地的南面为滨江绿化带和珠江。方案构思为“圆润双砾”，一大一小一黑一白的两块奇石安然矗立于珠江江畔。建筑造型组合形成了一个有视觉冲击力的形体，犹如鬼斧神工之作，天造地设而生，这是自然生长的大地建筑，极具未来感。

“大石头”为1800座的歌剧场及其配套设备用房；“小石头”为400座的多功能厅及西餐厅；南部的“草坡”是公共配套设施用房；地下室用作停车库、设备用房、部分

Guangzhou Opera House is located in the west of the intersection of the new city axis and the north bank of the Pearl River. To the east of the axis are Guangzhou Library and Guangdong Province Museum. To the north of the site is Guangzhou Second Children's Palace. To the south of the site are the riverside green belt and the Pearl River. According to the conception, two remarkable rocks, large and small, black and white, stand on the bank of the Pearl River, which conveys the meaning of roundness and fullness. Their architectural shapes combined constitute a physical body of visual impact as if built by the spirits. They are buildings naturally growing out of the ground, reflecting a strong futuristic style.

"Big Rock" refers to the Opera House with a capacity of 1,800 seats and its supporting equipment room. "Small Rock" refers to the multipurpose hall

化妆间及舞台机械设备的台仓。位于两块"石头"和"山丘"之间的首层部分为架空层，与相邻的水面和草坡共同构成一个可供公众开展文化艺术活动的开放空间。

外立面幕墙材料以天然花岗石为主，通过"大石头"和"小石头"在体形、色彩及质感上的对比、呼应，表达了自然粗犷和舒展飘逸的建筑空间效果。几条沿大小石头表面环绕的形态自由的玻璃带保证了室内公共空间获得极佳的珠江景观和自然光线。

奖项：

广州市 2010 年度优秀工程勘察设计一等奖；
广州市 2010 年度优秀结构专业设计一等奖；
广州市 2010 年度优秀建筑环境与设备专业设计一等奖；
广东省 2011 年度优秀工程勘察设计一等奖；
广东省 2011 年度优秀结构专业设计一等奖；
广东省 2011 年度优秀建筑环境与设备专业设计二等奖；
全国 2011 年度优秀工程勘察设计一等奖；
全国 2011 年度优秀结构专业设计二等奖；
全国 2011 年度优秀建筑环境与设备专业设计三等奖；
2013 年香港建筑师学会两岸四地建筑设计大奖卓越奖。

with a capacity of 400 seats and western restaurant. On the southern lawn slope lies the room for public supporting facilities. Basement will be used as garage, equipment room, dressing room and stage machinery room. The first floor between the two "rocks" and the "hill" is built on stilts, which together with the adjacent water surface and lawn slope provides a public open space for cultural and artistic activities.

The exterior facade is mainly built with natural granite. The use of contrast and complement in size, color and texture between the "Big Rock" and the "Small Rock" achieves architectural spatial effects that feature natural roughness, unrestraint and elegance. The glass belt along the surface of these rocks not only brings natural light to indoor public space, but also provides an excellent view of the Pear River.

Awards:

1st Prize for Construction Survey Design awarded by Guangzhou Municipal Government in 2010.
1st Prize for Structural Design awarded by Guangzhou Municipal Government in 2010.
1st Prize for Architectural Environment and Equipment Design awarded by Guangzhou Municipal Government in 2010.
1st Prize for Construction Survey Design awarded by Guangdong Provincial Government in 2011.
1st Prize for Structural Design awarded by Guangdong Provincial Government in 2011.
1st Prize for Architectural Environment and Equipment Design awarded by Guangdong Provincial Government in 2011.
1st Prize for National Construction Survey Design in 2011.
2nd Prize for National Excellent Design n 2011.
3rd Prize for National Architectural Environment and Equipment Design in 2011.
Outstanding Prize for Cross-Strait (Taiwan, Hong Kong, Macau and Chinese Mainland) Architectural Design Award by The Hong Kong Institute of Architects in 2013.

ifp

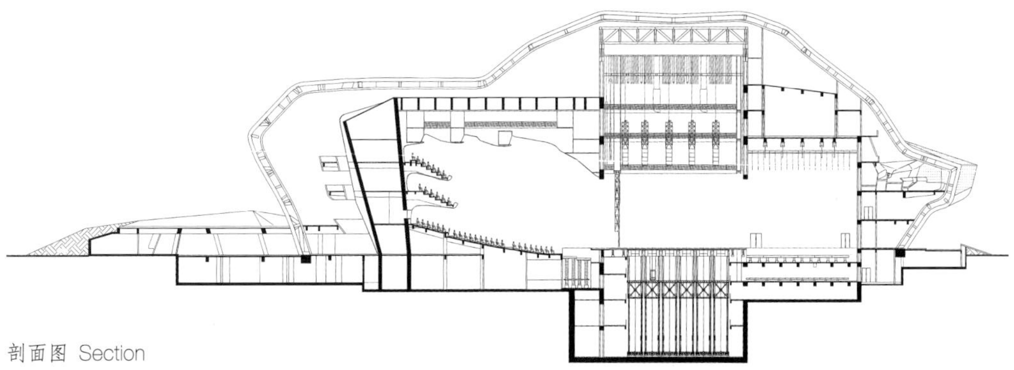

剖面图 Section

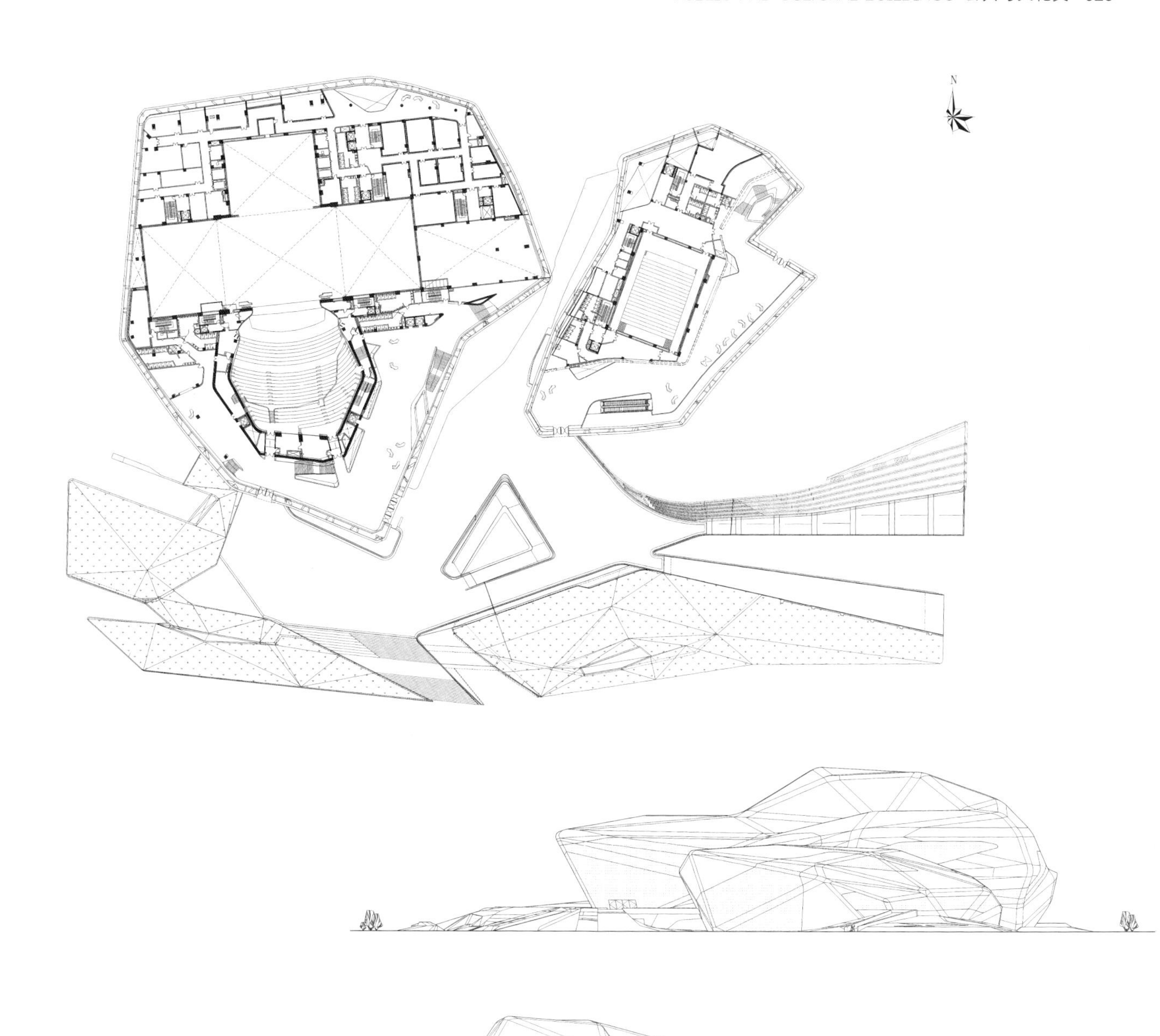

上至下：5m 标高平面图、东立面图、南立面图。
From top to bottom, Plan of 5 m Elevation, Illustration of the West Facade.

03 北京天桥艺术中心
Beijing Tianqiao Arts Centre

北京天桥艺术中心总建筑面积约7.4万平方米，包括1600座大剧场、1000座中剧场、400座小剧场、300座多功能厅四个特色剧场，艺术文化展览及艺术商业配套服务设施等。屋顶的曲线，建筑造型的柔美之魅，讲述着老北京的故事。屋顶的弧度，弯弯缓缓的线条的交织，构成了妙不可言的曲线美，形成了北京城市中轴的天际线。

天桥艺术中心将作为北京文化的标志性建筑、北京天桥文化的门户，勾起市民对传统演艺空间的回忆，对传统艺术的热爱。观演建筑的艺术文化理念与北京天桥的历史文化古韵相融，通过极具感染力的建筑艺术表现手法传达！

Beijing Tianqiao Arts Centre, with a GFA of 74,000 square meters, comprises four featured theatres, namely a 1600-seat Grand Theatre, a 1,000-seat Medium Theatre, a 400-seat Mini Theatre and a 300-seat Multi-purpose Hall, as well as art & culture exhibitions and commercial supporting facilities and services. The elegantly mellow building form with its curves of the roof relates the long standing history of the old Beijing city. The intriguing curvy contour of the building, as displayed in the elegant curvature of the roof and the interwoven gentle lines, forms the skyline of the city's central axis.

As a cultural landmark of Beijing and the icon of Tianqiao culture, Tianqiao Arts Centre evokes people's memory on traditional performance spaces and their passion for traditional art. The artistic and cultural concept of the theatrical building is mingled with the historic charm of Tian Qiao, and conveyed through expressive architectural approach full of artistic appeal.

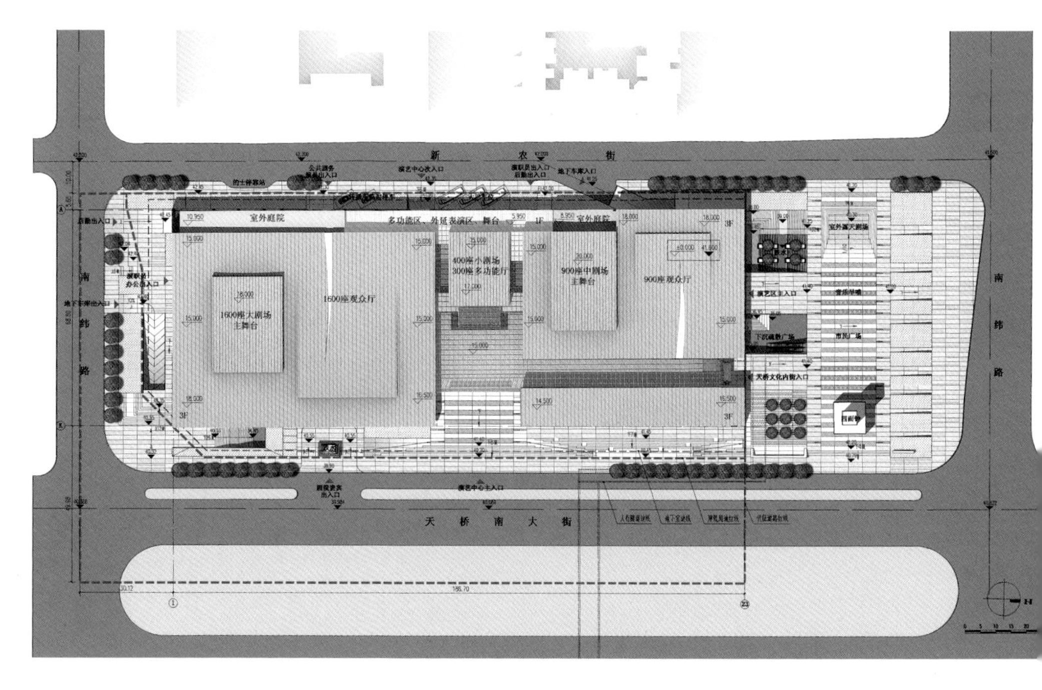
新 农 街
南 纬 路
天 桥 南 大 街
室外庭院
多功能区、外延表演区、舞台
1600座大剧场
主舞台
1600座观众厅
400座小剧场
300座多功能厅
900座中剧场
主舞台
900座观众厅
室外露天剧场
演艺中心次入口
演艺中心主入口
地下车库入口
天桥文化内街入口
下沉疏散广场
市民广场
演艺区主入口

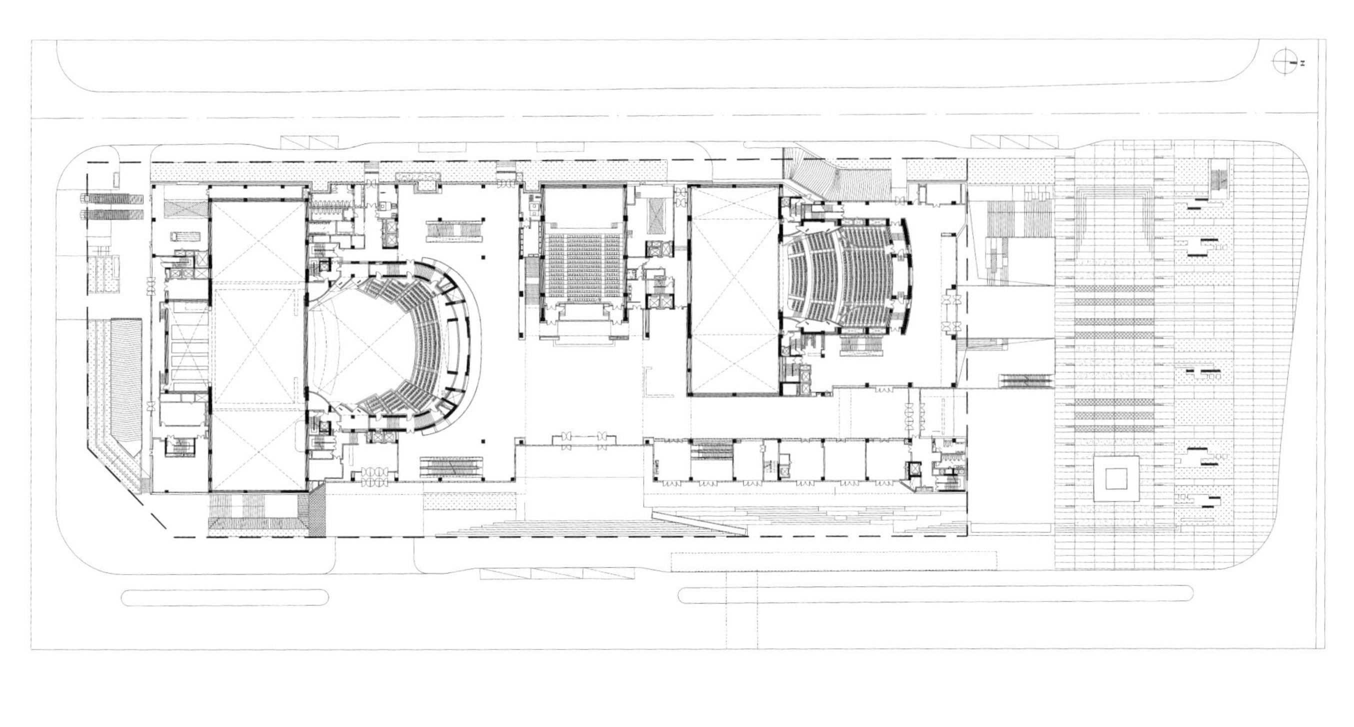

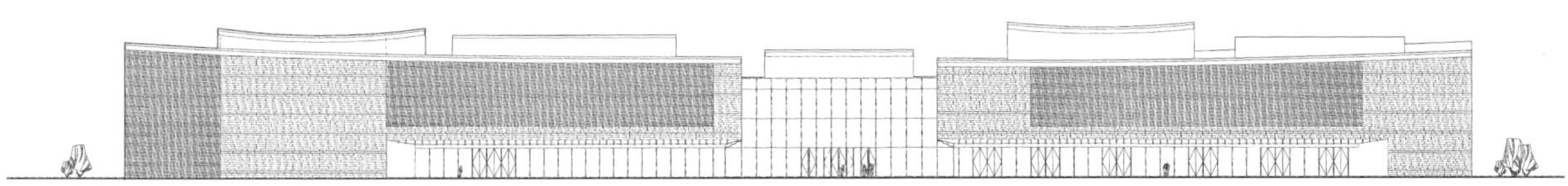

上至下：首层平面图、东立面图。
From Top, the First Plane, East Elevation.

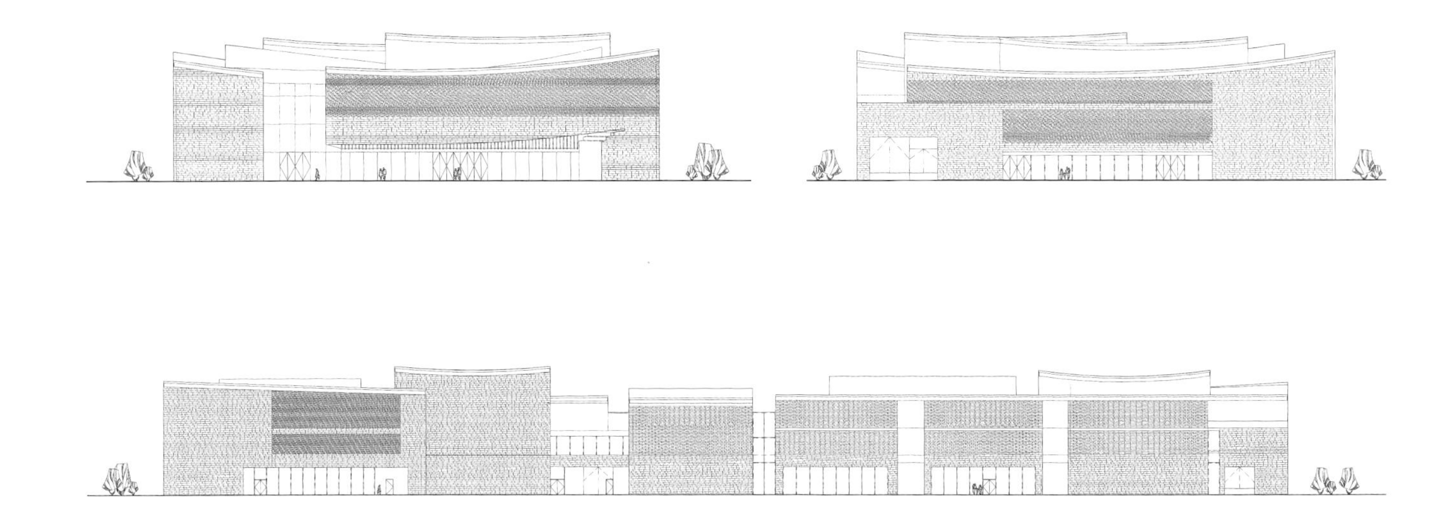

上至下：北立面、南立面、西立面图。
From Top, North Elevation, South Elevation, West Elevation.

04 长沙梅溪湖国际文化艺术中心
（与扎哈·哈迪德建筑师事务所合作设计）

Changsha Meixi Lake International Culture & Art Centre
(In collaboration with Zaha Hadid Architects)

长沙梅溪湖国际文化艺术中心，中标的英国扎哈·哈迪德建筑师事务所选择我院承担国内配合设计。

梅溪湖国际文化艺术中心包含了一系列独特的城市节点和空间：一个大剧院、一个现代艺术馆、一个小剧场和提供餐饮及其他配套的商业设施。各个建筑单体的形态与其功能相呼应。这种独立的有个性的形式强烈地显示着城市的易读性以及明确的方向性，营造了一个充满活力的城市方案，建立了一个有机的秩序和自然的优雅性。

Zaha Hadid Architects, the bid winner of the bidding for Changsha Meixi Lake International Culture & Art Centre, selected us as local design partner to work with them.

The Centre includes a series of unique urban nodes and spaces, i.e. a grand theatre, a modern art gallery, a mini theatre, catering services and other commercial supporting facilities. The form of each single building responds to the corresponding function. This stand-alone and individualized form strongly indicates the readability of the city and orientation to visitors, significantly contributing to an energetic urban plan in organic order and natural elegance.

05 武汉万达中央文化区秀场与六星酒店
Wanda Central Cultural Zone Theatre and Six-Star Hotel, Wuhan

武汉中央文化区位于武昌区东湖和沙湖之间，是武汉市核心地段。项目规划区域约1.8平方公里，总建筑面积300多万平方米，包括商业街、万达广场、酒店、写字楼、秀场、室内主题公园、游艇码头及居住区等，是万达集团投资数百亿元人民币倾力打造的以文化为核心，兼具旅游、商业、商务、居住功能的世界级文化旅游项目。

秀场建筑设计源自中国传统红灯笼造型，并采用现代手法与技术手段进行表现，以达到传统元素与现代技术的高度融合。红灯笼造型轻盈剔透的结构美可以通过三层挑高的大堂向上一览无遗，还能结合夜景照明进行完美展示。红灯笼与裙房屋顶衔接处的处理手法神似中国传统建筑屋顶，裙房柱廊墙壁上装饰有中国传统纹样的雕刻。

Wuhan Central Cultural Zone is prominently located at downtown between East Lake and Shahu Lake in Wuchang District. The planned project area is about 1.8 square kilometers with a GFA of about over 3 million square meters, including shopping streets, Wanda Plaza, hotels, office buildings, theatre, indoor theme parks, marina and residences. Wanda Group has invested billions of RMB to complete this world's top-grade cultural tourism project centering on culture and integrating tourism, business, commerce and residence.

The theatre design is inspired by the traditional Chinese red lanterns and impeccably combines traditional elements and modern technologies through modern technical means. The graceful clarity of the red lantern structure is perceived readily through the three-floor-high lobby, while the night lighting even reinforces the glamour. The connection between the "Red Lantern" and the roof of podiums is handled in a way that is closely similar to the roof of traditional Chinese architecture. The walls of the colonnade in the podiums are carved with traditional Chinese artistic patterns.

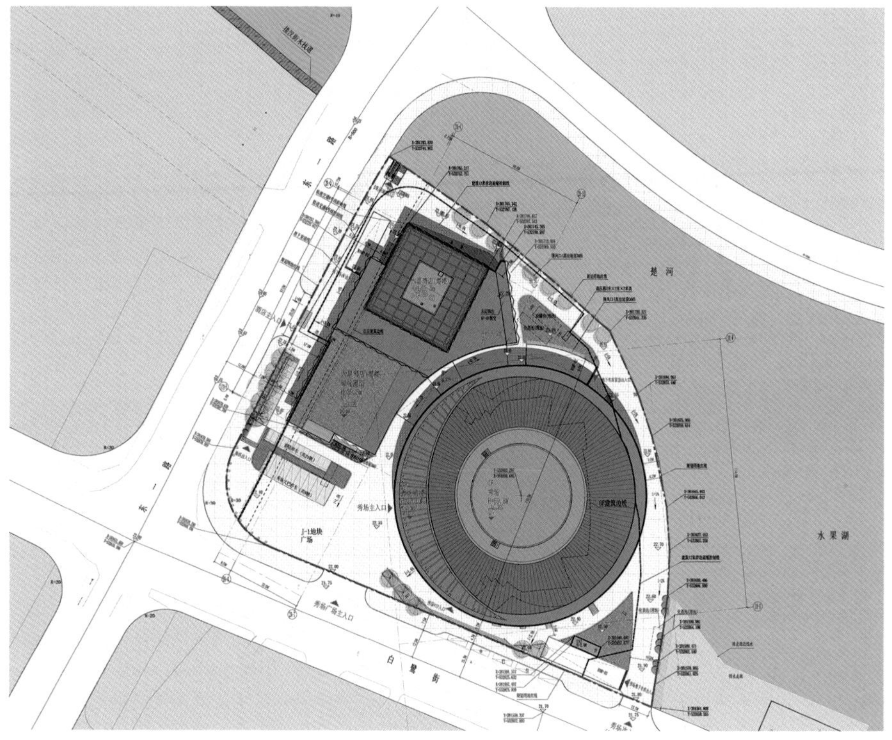
楚河
秀场主入口
J-1地块
广场
水果湖
秀场广场主入口
白鹭街
6F建筑边线

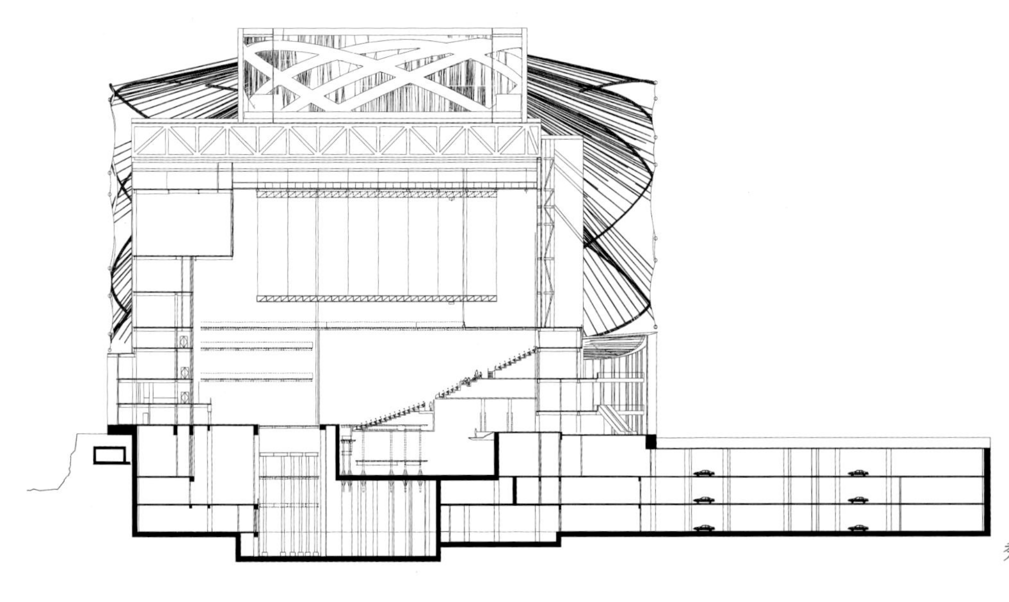

秀场剖面图 Theatre Section

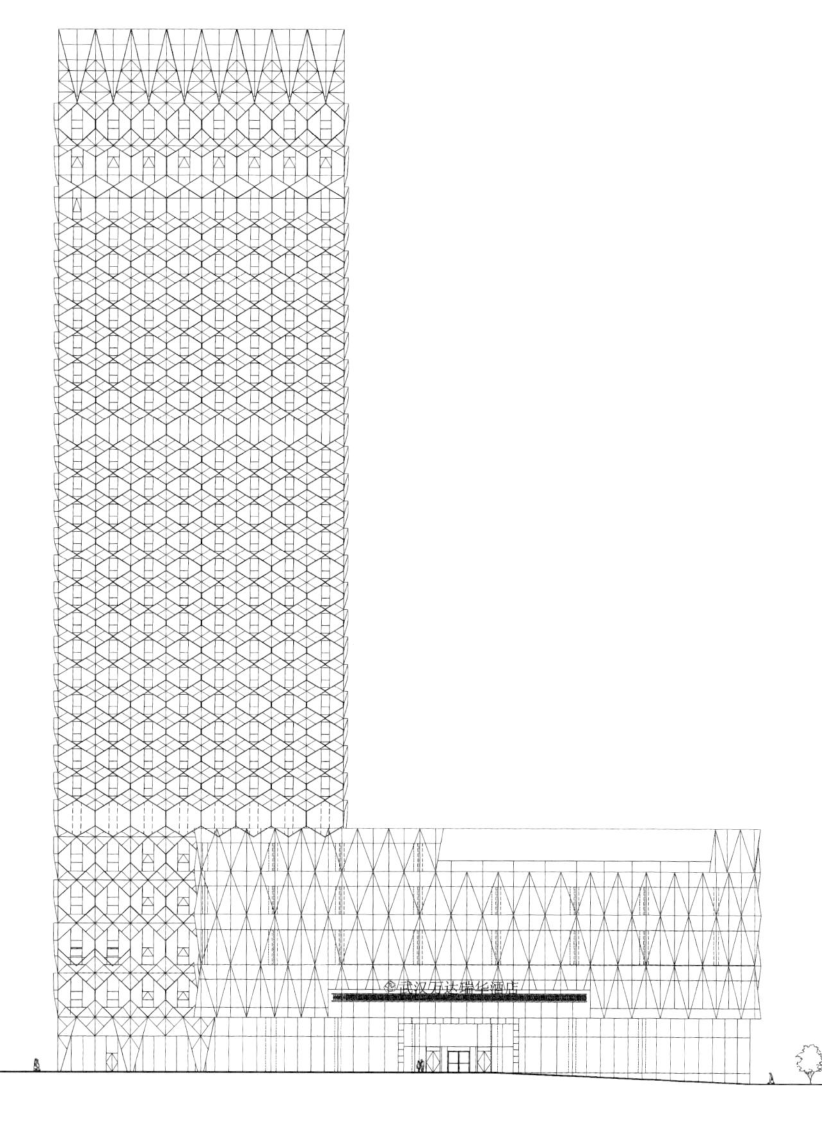

酒店立面图 Hotel Elevation

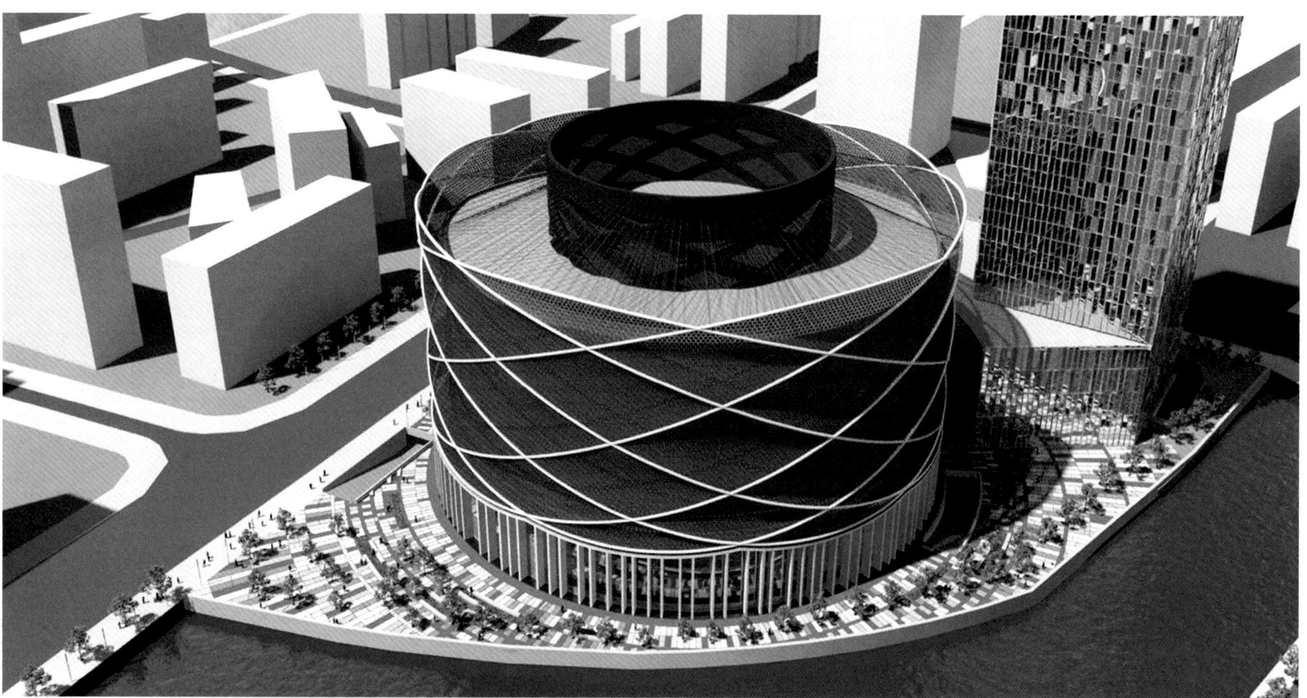

06 珠海十字门国际会议中心（与英国 RMJM 建筑事务所合作设计）

Zhuhai Shi Zi Men International Convention Centre (In Collaboration with RMJM Architects, UK)

国际会议中心位于珠海市香洲区湾仔，南湾大道东南侧，面向大海，北、西、南面分别与商业零售、国际展览中心及标志性塔楼相邻。其中，多功能厅 1 为一个 1200 座的观演厅，以演出歌剧、舞剧为主；多功能厅 2 是一个 2000 平方米的可弹性划分的多功能会议厅，内部空间可划分为 3 个独立使用的分厅；多功能厅 3 为 800 座纯音乐厅，满足中型交响乐及室内乐演出的要求。

Situated in Wanzai, Xiangzhou District, Zhuhai, the Centre, facing the South China Sea, is to the southeast of Nan Wan Boulevard and borders on the retails on the north, International Exhibition Centre on the west and a representative tower building on the south. Multi-purpose Hall I in the Center is a 1,200-seat theatre hall to accommodate operas and dance dramas. Multi-purpose Hall 2 in a floorage of 2,000 square meters is a flexibly-dividable multi-purpose conference hall which can be sub-divided into 3 independent halls. Multi-purpose Hall 3 is an 800-seat concert hall to house medium-sized symphonies and chamber music.

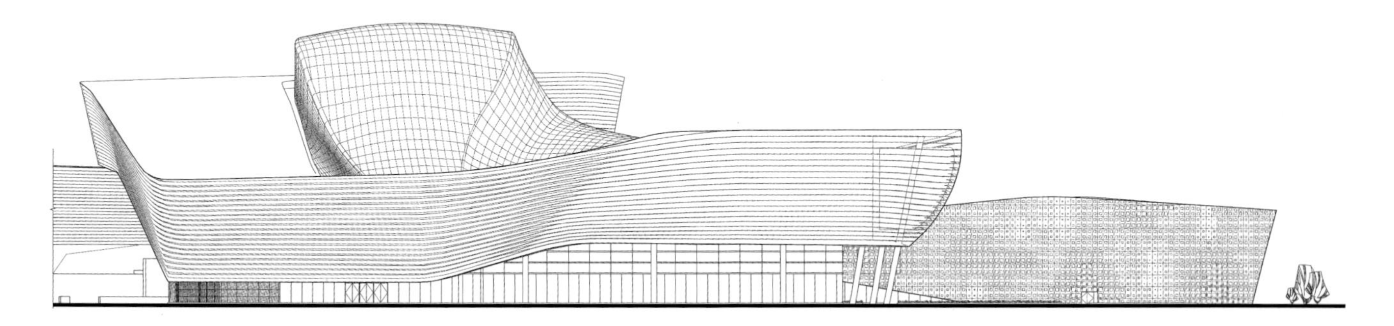

右上：二层平面图 Upper Right: Second Floor Plan

左下：南立面图 Lower Left: South Elevation

右下：东立面图 Lower Right: East Elevation

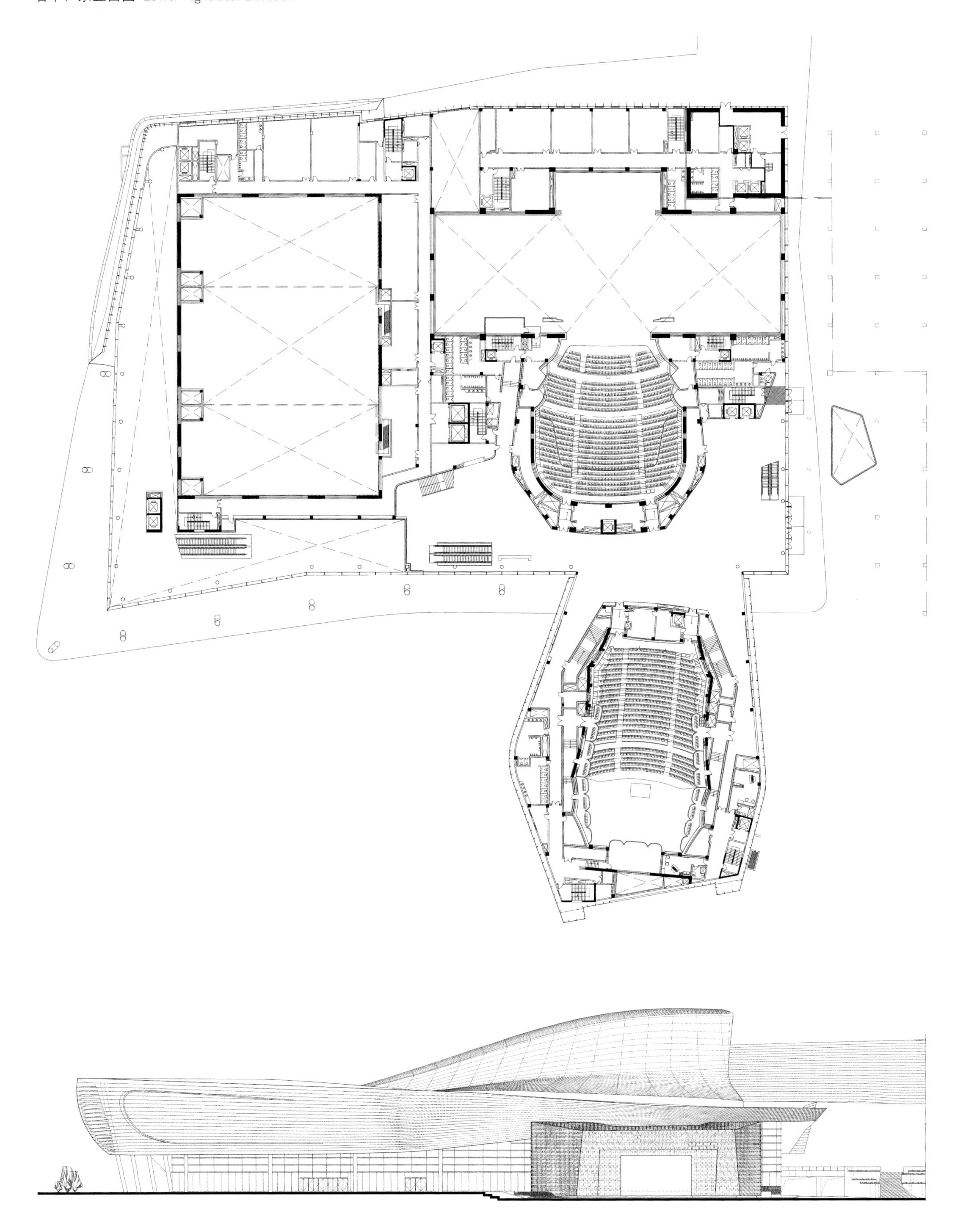

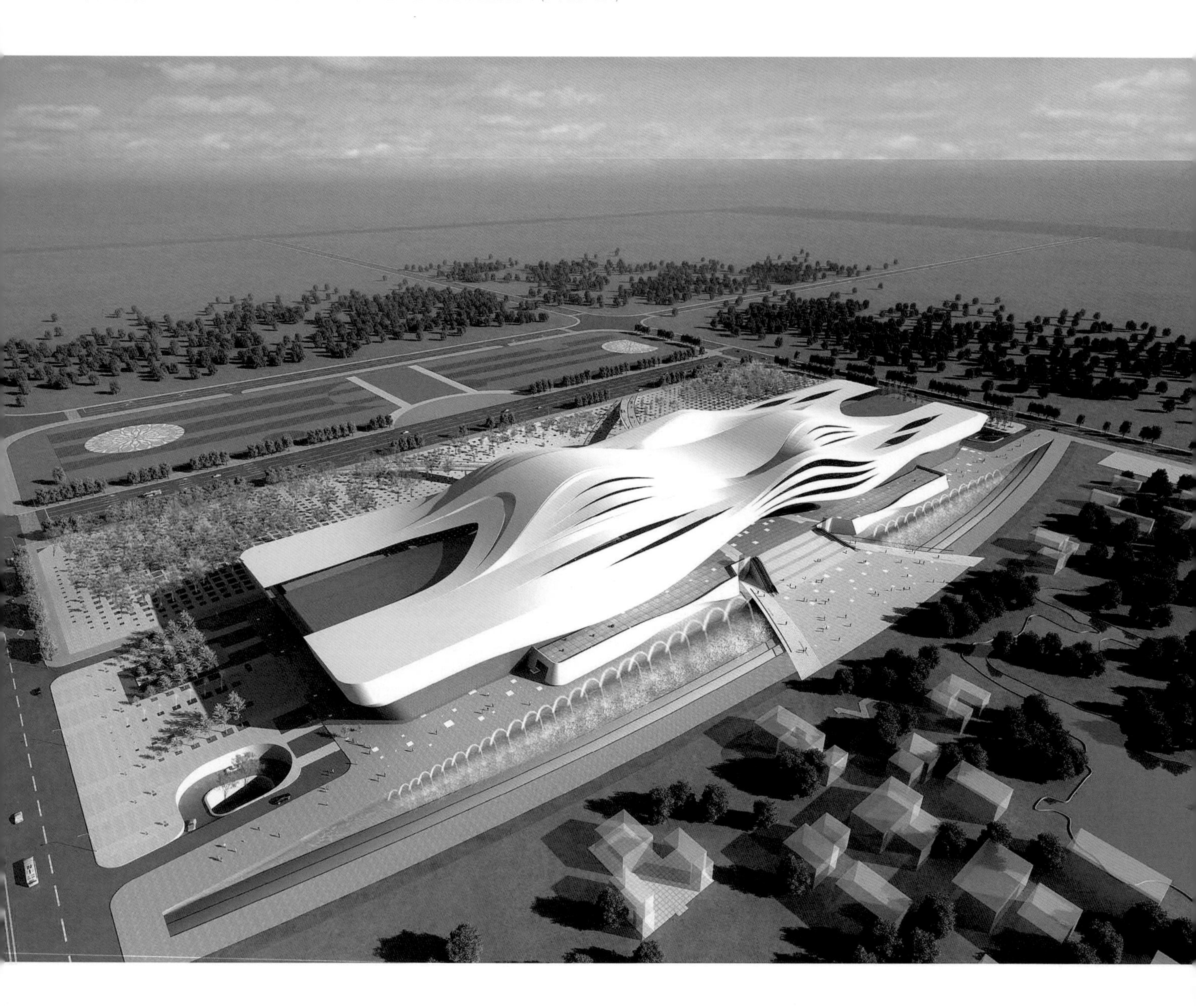

07 廊坊大剧院
Langfang Grand Theatre

“梦廊坊”文化产业园和谐世界主题园区位于廊坊的北部，东临和平路，北临新奥艺术大道，西临新开路，南侧与廊坊市文化中心融合，廊坊大剧院位于主题园区的内部。

在华北平原上，大地突然从这里隆起，在天空的覆盖下，一朵祥云飘落在隆起的绿色山丘上，伴随着天地交融，主题大剧院出现在人们的眼前。利用自然和戏剧两大主题打造了浑然天成、特点鲜明的建筑外观造型。

Harmonious World Theme Park of "Dream Langfang" Cultural Industrial Park, located in the north of Langfang, Hebei Province, borders on Heping Road on the east, Xin' ao Art Boulevard on the north, Xinkai Road on the west and Langfang Culture Centre on the south. Langfang Grand Theatre is situated within the Theme Park.

The flat land bulges here on the North China Plain. Right under the boundless firmament, a graceful "cloud", the Grand Theatre, lands on the bumped emerald hills emerges before people against a spectacular background view of the sky and the earth. The seemingly naturally bestowed and uniquely featured building form focuses on two themes: nature and dramas.

总体日景鸟瞰图

08 江门演艺中心
Jiangmen Performing Arts Centre

本方案取义于中国传统哲学“有生于无”、“阴阳和谐”、“和而不同”的观点，建筑体量由简单的方形分离出凹凸两个形体，两者之间分离后，其中的空间即为广场空间。这个整体关系中既有两实体（演艺板块与文化板块）之间的对话，又有虚与实（广场与建筑）的对话，表现了“阴阳和谐，生生不息”。

在建筑造型及表皮应用上，演艺中心吸取了江门地方民俗文化中的“鹤山舞狮”、“陈山火龙”中的元素，以抽象的形式表达了江门舞龙和舞狮的民俗文化。剧院部分

The design is inspired by traditional Chinese philosophical thinking like "being comes from nonbeing", "a balance of *Yin* and *Yang*" and "harmony without uniformity". The building form is created by splitting one plain cube into two volumes, one concave and one protruding, while the space between the two volumes is left open as a square. This way the Project comprises not only the dialogue between two solid volumes (the performing part and the cultural part), but also that between the solid and virtual volume (the building and square), echoing to the design philosophy of "everlasting balance between Yin and Yang".

The building shape and skin abstractly convey the folk culture of the "lion dance" and "dragon dance" in Jiangmen by employing the local cultural elements from "Lion Dance of Heshan City" and "Dragon Dance of Chenshan Village".

立面单元取自"台山冲蒌编织"，利用编织纹理的通透性产生半透明的光影效果，既起到遮阳作用，又丰富了立面及室内空间效果。

奖项：

2011 年度广东省注册建筑师协会优秀建筑创作奖。

The certain part of the theatre facade derives its inspiration from the weaving art of Chonglou Town, Taishan City. The permeability produced by the patterns of the weaving renders a translucent effect of light and shade, protecting the interior space from the sun and meanwhile enriching the facade and the interior space.

Award:

Excellent Architecture Creation Award of Guangdong Chapter of Association of Chinese Registered Architects (ACRAGD), 2011.

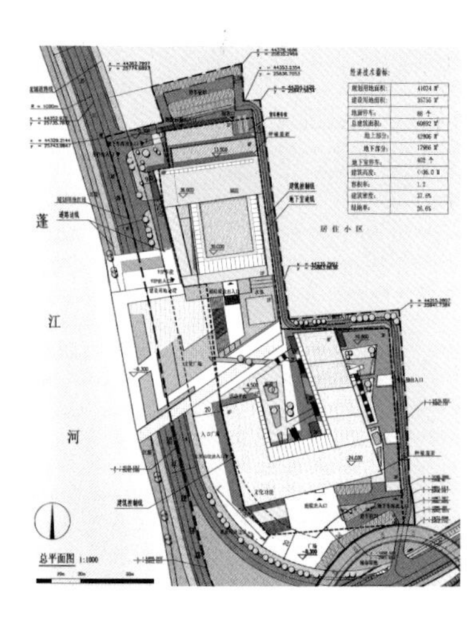

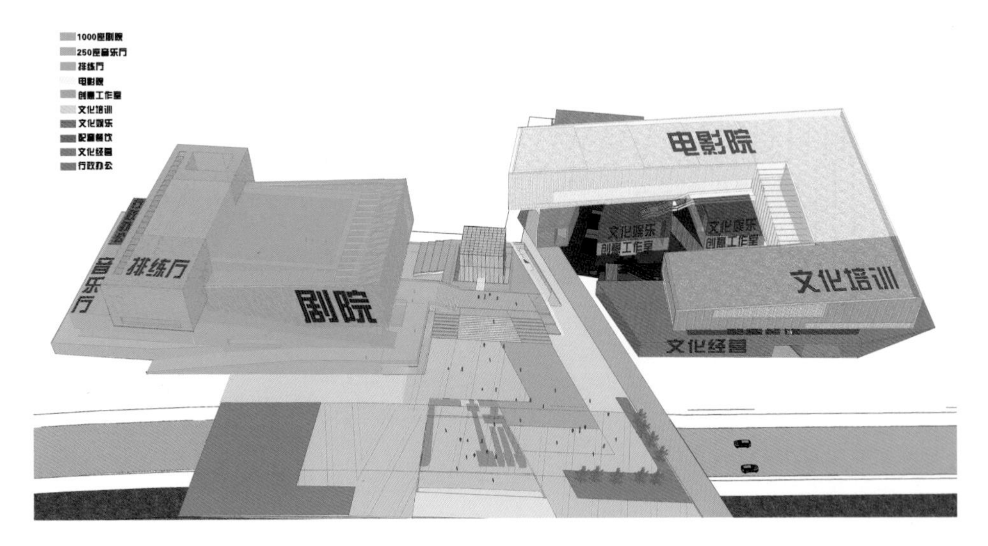
1000座剧院
250座音乐厅
排练厅
电影院
创意工作室
文化培训
文化娱乐
配套餐饮
文化经营
行政办公
电影院
文化娱乐
创意工作室
文化娱乐
创意工作室
文化培训
文化经营
音乐厅
排练厅
剧院
广场

JIANHAI

09

长春净月潭保利大剧院
Poly Grand Theatre, Jingyuetan, Changchun

长春净月潭保利大剧院总建筑面积约为5.5万平方米（含地下室面积），包括1400座的大剧院、400座的多功能厅、为整个项目配套的餐厅、剧务用房、演出用房、公共剧务用房、通用设备用房、行政用房、业务办公用房、后勤用房、停车库、公共配套设施用房等部分及完善的艺术品商店、酒店等剧院配套设施。

保利大剧院将建设成为一个功能齐全、视听条件优良、技术先进可行、设施完善、完全适用、经济合理的国家级剧院。它是长春政府关注的重点项目，建成后将作为吉林省的标志性建筑之一，与周边项目形成综合的城市公共区域。

Poly Grand Theatre, Jingyuetan, Changchun, with a GFA of about 55,000 square meters (basement area included), is made up of a 1,400-seat grand theatre, a 400-seat multi-purpose hall, a restaurant serving the entire Project, stage management offices, performance rooms, public stage management offices, general mechanical rooms, administration offices, business offices, BOH rooms, garages, public supporting facilities and well-equipped supporting facilities for artwork business hotels and the theatre.

Poly Grand Theatre aims to be a fully-functional and well-equipped national-level theatre boasting up-to-date technical systems (in particular the A/V system) and reasonable construction and operation costs. As one of the key projects of Changchun city government, the Theatre will become a landmark of Jilin Province and create a comprehensive urban public area together with the other projects in the periphery.

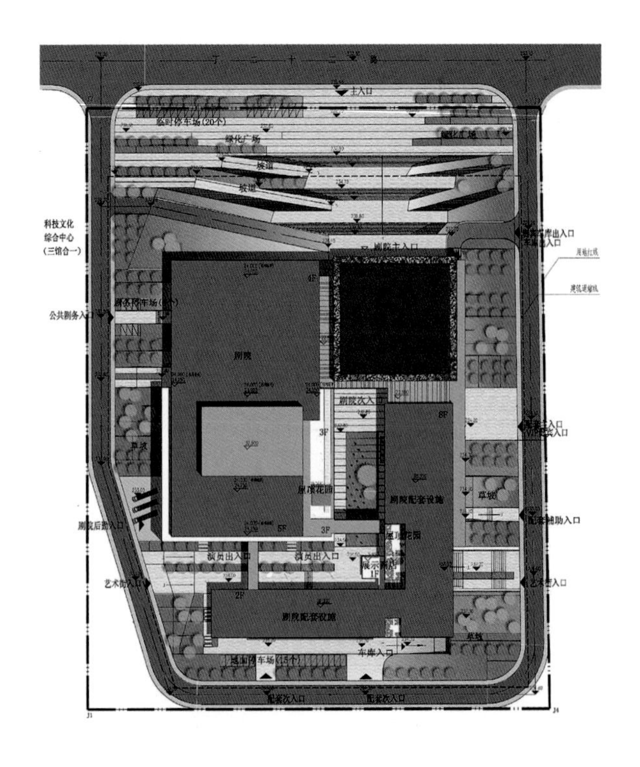

10 汶川县禹羌博物馆
Wenchuan County Yu Qiang Museum

设计将博物馆建筑主体布置在用地北侧及东侧，呈"L"形布置，东侧的建筑主体沿威州镇中轴线布置，围合出的空间成了供市民活动的开放型禹羌广场，极大地减弱了周边建筑对该综合体建筑的干扰与影响，并通过拾级而上的平台，营造出纪念建筑的特有氛围。

博物馆的建筑形体以简洁、厚重、雄浑、伟岸、抽象的形式，屹立于汶川县城，刚劲有力。在禹羌广场之上，设计了体现羌族传统和文化特色的建筑——碉楼。羌族碉楼被喻为东方金字塔，是一个坚固的文化符号，它记录了羌民族勤劳、艰辛、纯朴的历史，也展示了他们独特、精湛的建筑艺术，使他们在中国建筑史上独树一帜。

奖项：

广州市2010年度优秀工程勘察设计二等奖。

The main museum building in "L"-shape layout is placed in the north and east of the site. The east wing extends along the central axis of Weizhou Town. The open space thus enclosed, the Yu Qiang Square, offers ideal place for the public activities of citizens, meanwhile, greatly mitigates the interference and adverse effect of surrounding buildings over the museum complex. The platform with ascending stairs helps foster the peculiar atmosphere of a monumental building.

The museum building stands boldly and powerfully with concise, profound, magnificent yet abstract form in Wenchuan County. Moreover, watchtowers, the building form that can best reflect the tradition and culture of Qiang people, are designed on Yu Qiang Square. The watchtower, reputed as "Oriental Pyramids", is a long-standing cultural symbol. It records the laborious history of the plain Qiang people, depicts their uniquely exquisite architectural art and adds an original typology to the Chinese architectural history.

Award:

The Second Prize for Excellent Engineering Exploration and Design of Guangzhou, 2010.

川博物馆

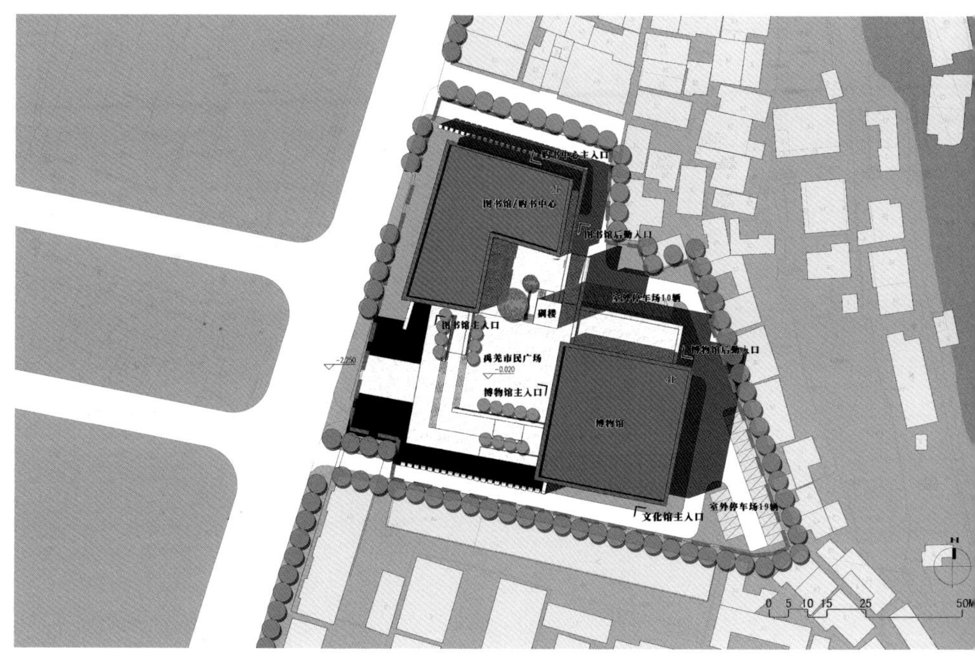
购书中心主入口
图书馆/购书中心
图书馆后勤入口
室外停车场10辆
钢楼
图书馆主入口
禹羌市民广场
-0.020
-2.250
博物馆主入口
博物馆后勤入口
博物馆
文化馆主入口
室外停车场19辆
0 5 10 15 25 50M

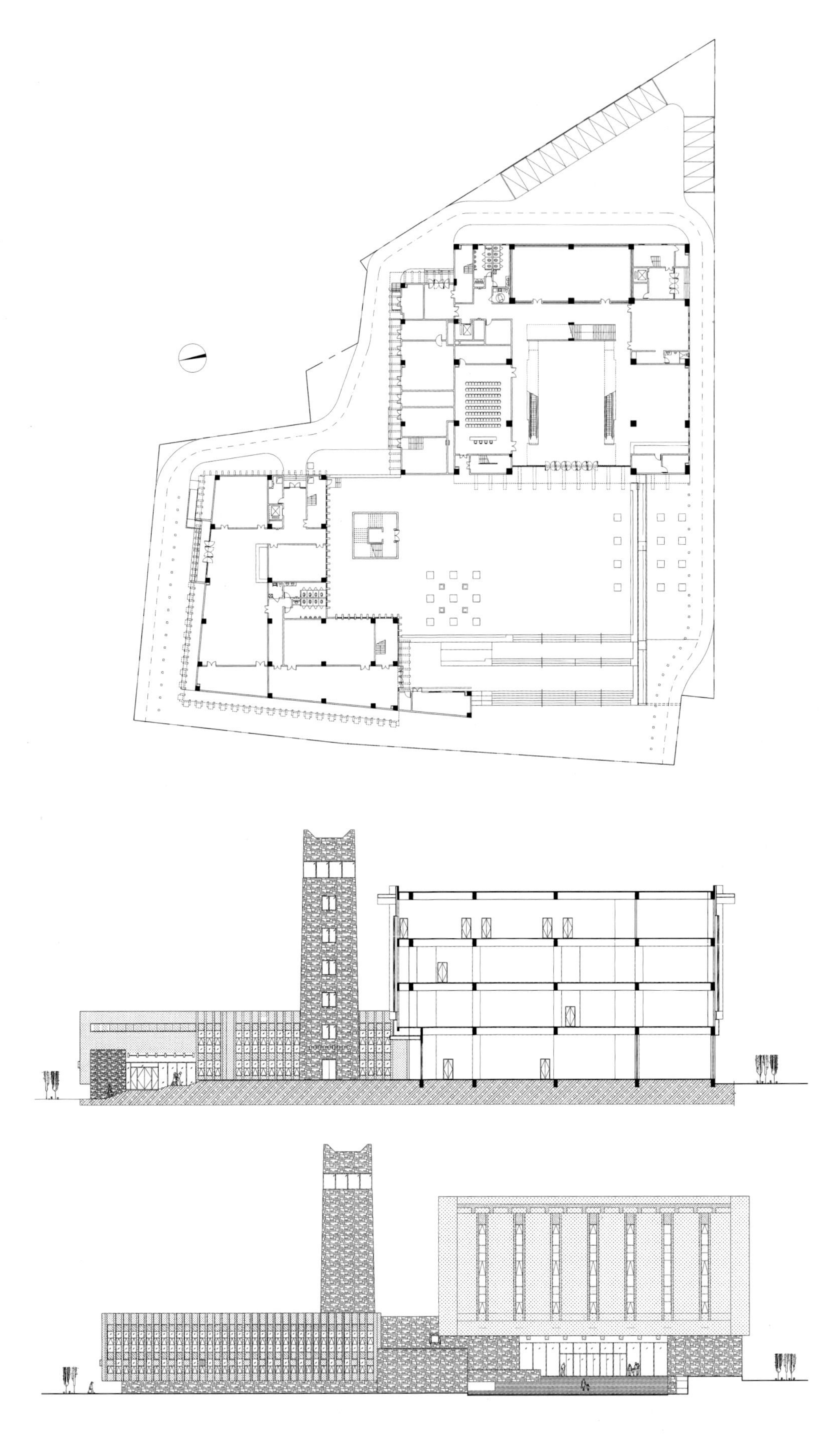

上至下：首层平面图、剖面图、立面图。
Top Down: Ground Floor Plan/ Section / Elevation.

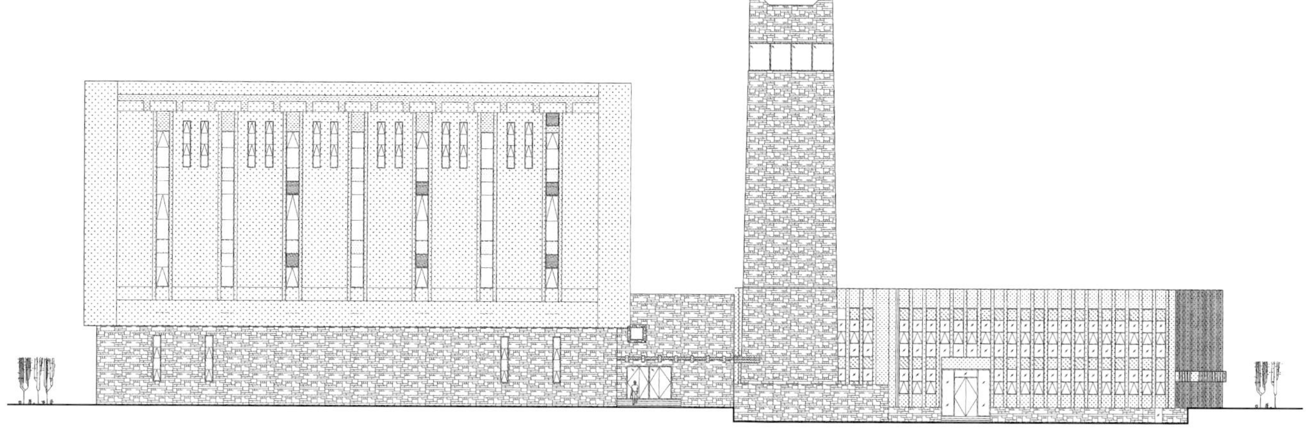

立面图
Elevation

汶川博物馆

汶川博物馆

11 汶川避灾广场
Wenchuan County Refuge Square

汶川避灾广场建设规划用地位于县城新建体育馆与阿坝州迎宾馆之间，南邻县政府办公楼，北靠岷江，用地面积为 21620 平方米。避灾广场设计旨在成为用作县城紧急避灾场所，突出时代特色和地域特色，反映城市和场地历史，满足集散、市民休闲、娱乐及举行仪式等需求的综合性城市开放空间。

本方案利用曲线形的纹理统筹整个场地。曲线形的纹理就像水流一般，起源于岷江，并向着这片土地上聚居千年的羌族聚落延伸，表达对孕育了羌族人民的"母亲河"——岷江的赞美，水流状的曲线围合出供市民休闲活动的圆形场地，圆形场地代表在水流中打磨，最后变得圆润的石头。场地中央设计了一条南北贯穿的中轴线，在中轴一端，即入口处，摆放主题雕塑，雕塑是一群小朋友围着援建志愿者跳舞的欢快场景，表达了灾后重建下汶川的美好未来。

奖项：

广州市 2010 年度优秀工程勘察设计二等奖。

The proposed site of Wenchuan County Refuge Square, between the newly-built stadium of the County and Aba Hotel, borders on the office building of County government on the south and the Minjiang River on the north. With a site area of 21,620 square meters, the project is designed to be a refuge for emergencies in the County. Bearing the modernism and local features, the Square is a reflection of the history of the County and the site. As a comprehensive urban open space, the Square aims to meet the needs of public gatherings, recreation, entertainment and ritual practices of residents of the County.

The entire site is designed in curve-shaped pattern which is akin to currents. This pattern, a eulogy of the Minjiang River, the mother river feeding Qiang people, starts from the Minjiang River and extends itself to the Qiang people's settlement with a history of a thousand years. The current-shaped curves enclose a round place for the recreation of citizens. The round place is symbolic of a mellow stone finally washed by waters. In the middle of the Square lies a south-north central axis, at one end of which the theme statue is displayed. The statue, symbolic of the promising future of the earthquake-stricken Wenchuan, delineates a jubilant scene where children are dancing around the volunteers aiding the rebuilding of the County.

Award:

The Second Prize for Excellent Engineering Exploration and Design of Guangzhou, 2010.

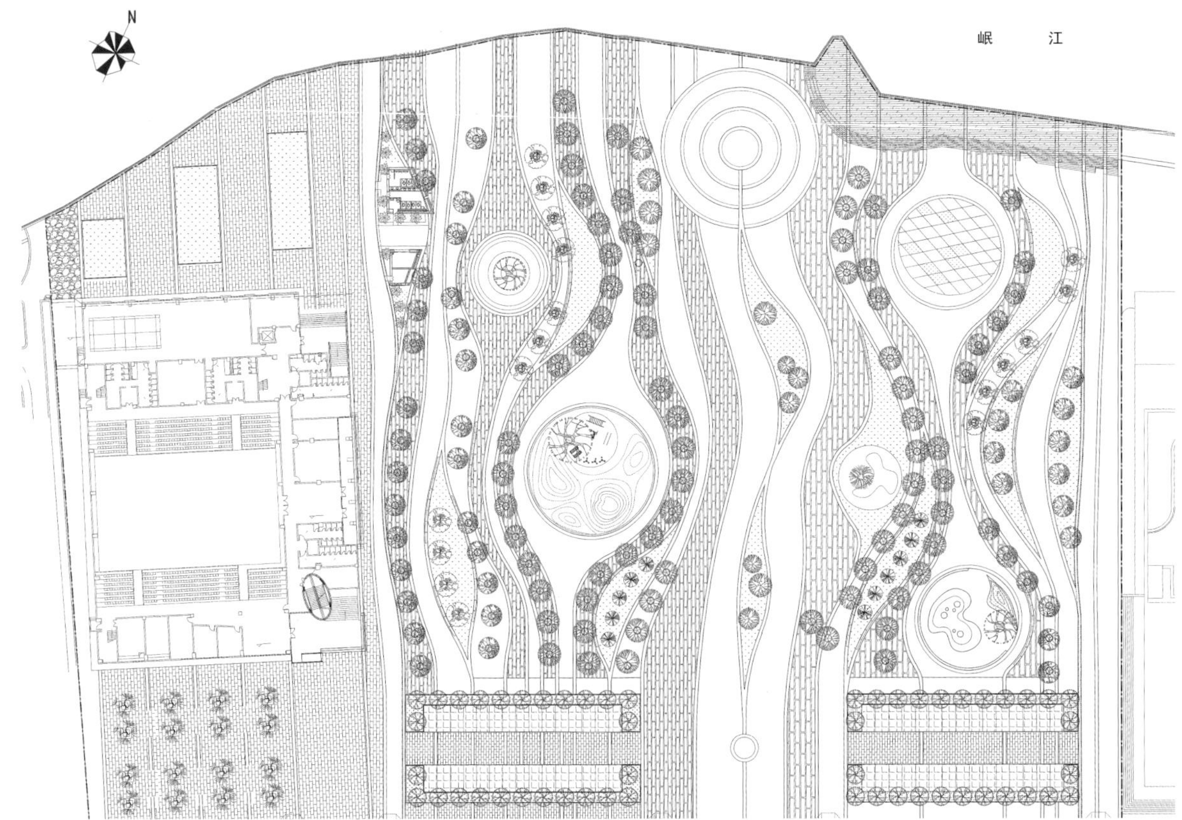
N
岷　　江

广州援建

12 武汉中山舰博物馆
"Zhongshan" Warship Museum, Wuhan

“中山”舰是中国近代海军史上一艘著名的军舰，它经历曲折的航程，曾创立过不朽的功绩，是具有世界影响的中华民族的“记忆”载体，具有极大的历史意义和文物价值。中山舰博物馆（核心）区位于武汉市江夏区金口古镇，牛头山山脊以南，金鸡山以北，西至长江江滩，东至金鸡湖湖边线，占地468.59亩，中山舰博物馆建筑面积为10164平方米。

建筑以简洁、厚重、抽象的形式屹立于湖边，形体刚劲有力。建筑的纯净形态传递着自然山势的信息，并寓意着和过往历史的关联；而由圆形纪念大厅、梯形的舰体陈列厅、船型的建筑主体、刚正笔直的湖面栈道以及屋面大阶梯组合

"Zhongshan" Warship, a foremost warship on the China's modern navy history, weathered tough times and performed meritorious deeds. As the "memory" carrier of Chinese nation, "Zhongshan" Warship with world influence is a heritage of great historical and cultural significance. Located in Jinkou Ancient Town, Jiangxia District, Wuhan, the Museum (central part) borders on ridges of Niu Tou Mountain on the north, Jin Ji Mountain on the south, the marshland of the Yangtse River on the west and the lake line of Jin Ji Lake on the east, totaling a site area of 468.59 mu and a GFA of 10, 164 square meters.

The museum building stands boldly and powerfully with concise, profound yet abstract form beside the lake. Meanwhile, the purity of the building form convey the messages from the mountains and symbolizes the connection between the present and the past. Moreover, the splendidly

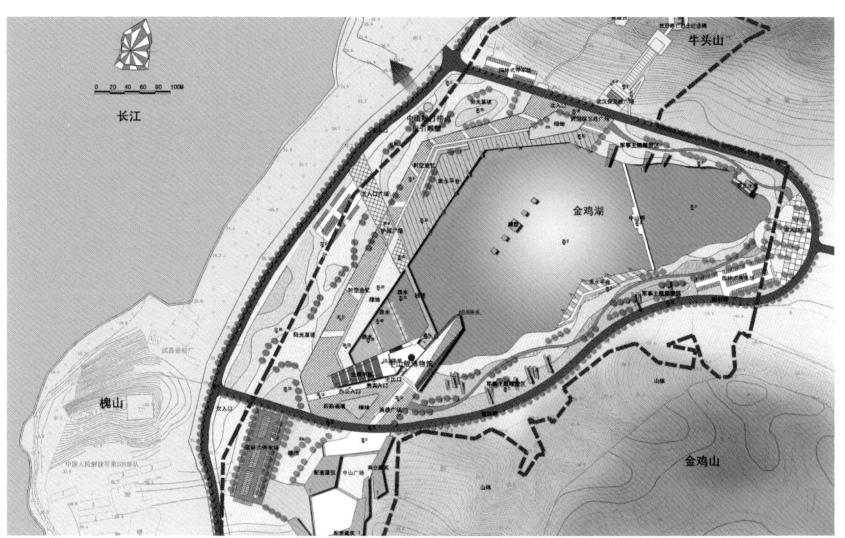

而成的形体，刚劲坚拔、自由舒展，犹如破水而出的战舰，同时通过材质的虚实对比和碰撞，塑造了一个英雄纪念碑的形象。

奖项：

广东省 2009 年度注册建筑师协会创作奖；
广州市 2010 年度优秀工程勘察设计一等奖；
广东省 2011 年度优秀工程勘察设计一等奖。

upstanding ensemble, composed of a round memorial hall, a trapezoid-shaped hull exhibition hall, the ship-shaped main building, as well as the structure made up by the perfectly straight pier on the lake and the large roofing flights, naturally extends itself with a striking resemblance with a real warship breaking the waters. The contrast and conflict between solid material and void enhance the Museum's presence as a "monument of the heroes".

Awards:

Creation Award of Guangdong Chapter of Association of Chinese Registered Architects (ACRAGD), 2009.
The First Prize for Excellent Engineering Exploration and Design of Guangzhou, 2010.
The First Prize for Excellent Engineering Exploration and Design of Guangdong Province, 2011.

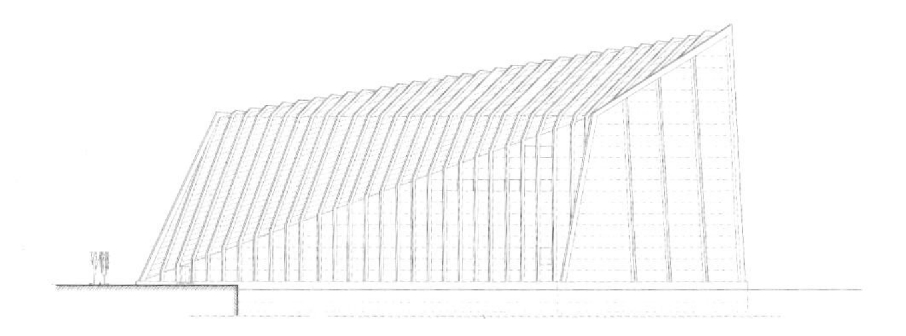

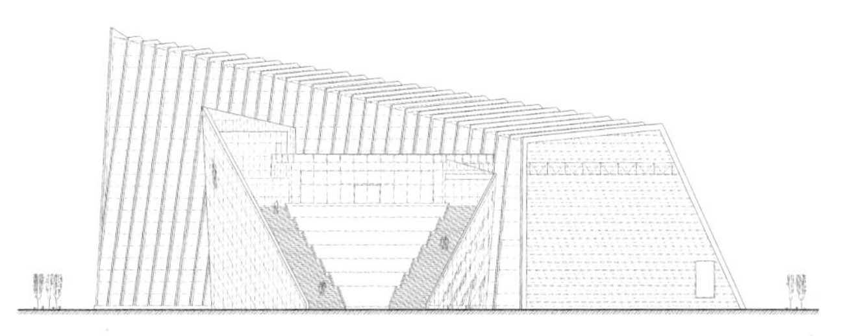

立面图
Elevation

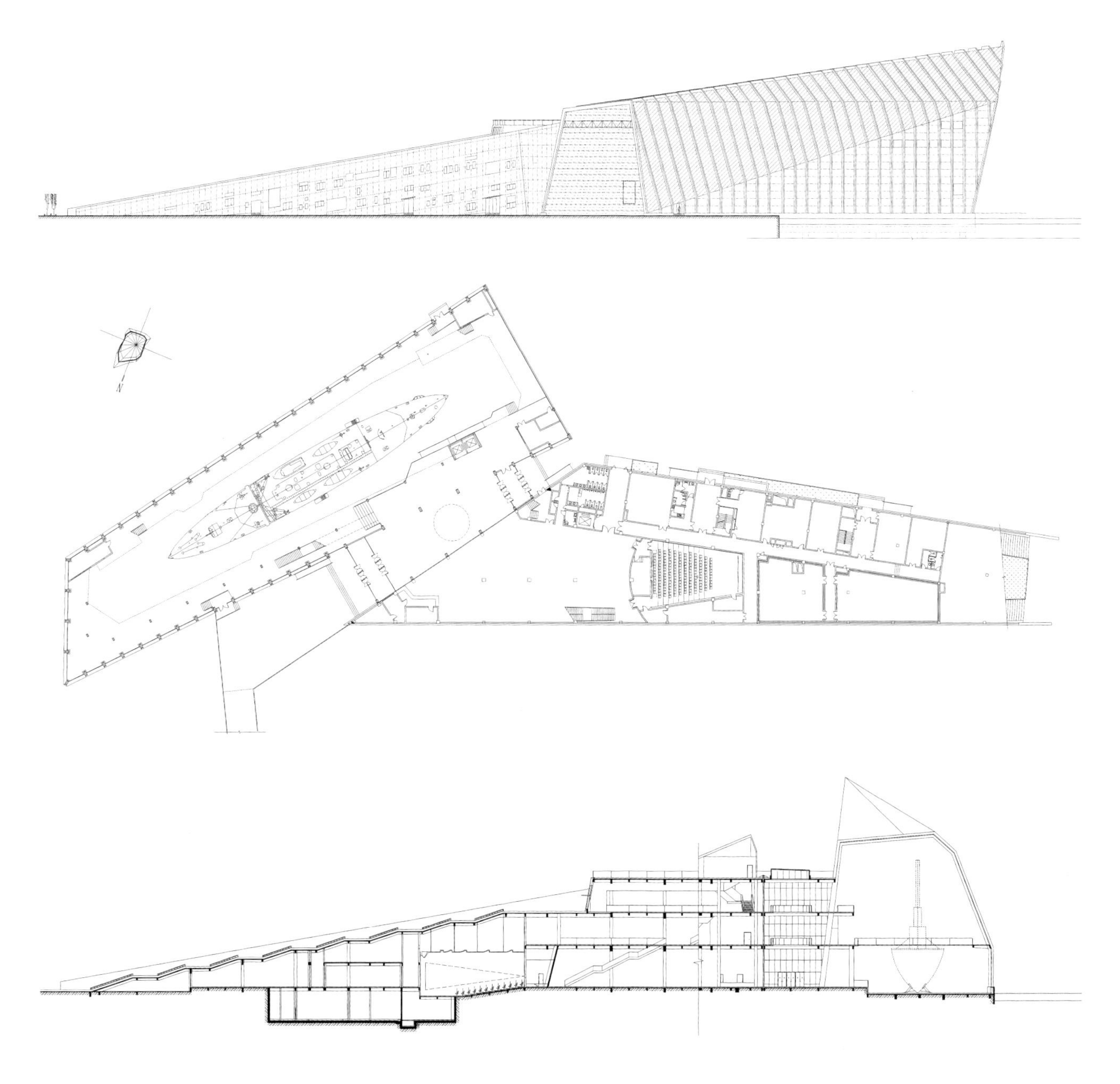

上至下：立面图、±0.000m标高平面图、剖面图。
Top Down: Elevation, ±0.000m Level Plan, Section.

13 广州辛亥革命纪念馆
Guangzhou Memorial Hall of the 1911 Revolution

辛亥革命纪念馆基地位于广州长洲岛的长洲镇，南邻珠江支流，西邻中山公园，东南面有低矮的山林，环境优美。

建筑形体构思源于一方顽石，一条"路"自南向北将顽石分为东、西两半，确定了总体的功能分布：西边部分为展览功能区，东边部分为后勤功能，"路"为参观主轴线。整体建筑如一方石块，平静地置于历史的长河之中。石块已被凿开，隐喻辛亥革命时期英

The project site, amid attractive surroundings and situated at Changzhou Town on Changzhou Island, Guangzhou, borders on a branch of the Pearl River on the south, Zhongshan Park on the west and low ranges of hills on the southeast.

The building form is inspired by a big stone. A south-north road "cuts" the stone into two parts: one on the east and the other on the west, which define the functional distribution of the Project. The west part is for exhibition and the east for BOH with the road as the main visiting axis. The entire building is like a stone peacefully residing in the river of the history. The chiseled stone is emblematic of the courage of starting the revolution and the painstaking fight of heroes during the 1911 Revolution.

烈们开天辟地的勇气和艰苦卓绝的历程。

插入石穴的"路"寓意"共和之路"，道路上随机陈列着辛亥英烈铸铜雕像，似在沿路而行。孙中山先生领导的辛亥革命在走向共和之路上前赴后继，百折不挠。这是一条求索之路，奋斗之路，牺牲之路。

奖项：

广州市 2012 年度优秀工程设计一等奖。

Alongside the road cut through the stone, symbolic of "Road to Republic", randomly lie bronze statues of the Revolution's heroes who have an air of walking along the road. The 1911 Revolution led by Dr. Sun Yat-sen underwent a series of arduousness yet unyielding pursuits for the Republic. This is a road of exploration, struggles and sacrifices.

Award:

The First Prize for Excellent Engineering Design of Guangzhou, 2012.

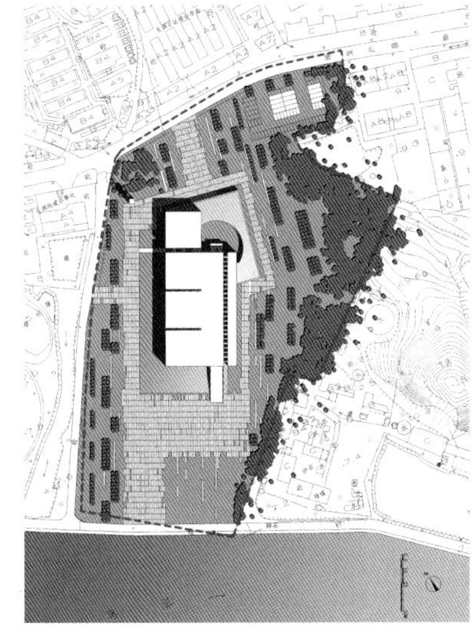

顽石开裂
Chiseled Stone

足迹
Footsteps

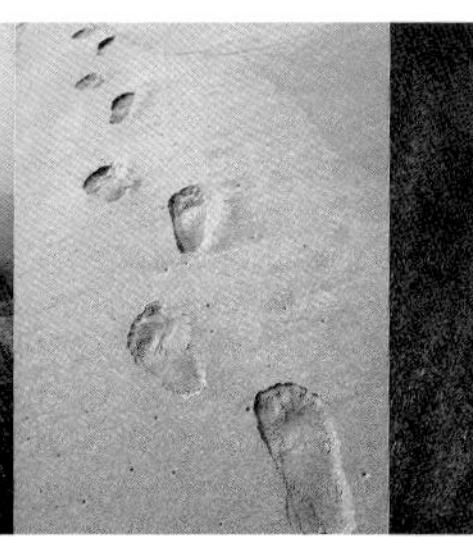

探索之路
Road of Exploration

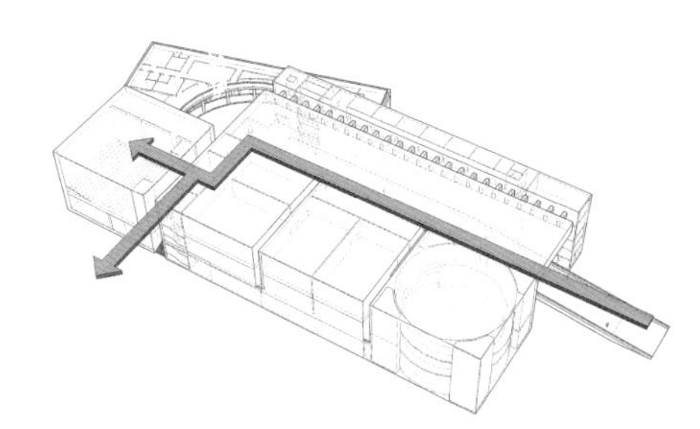

报告厅流线
Report Hall Route

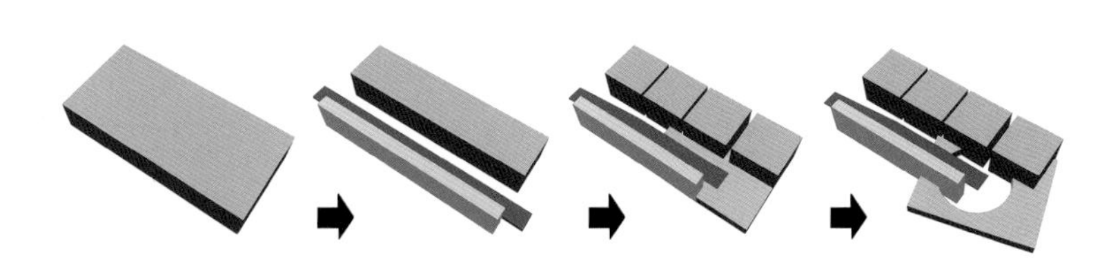

建筑形体分析
Analysis of Building Form

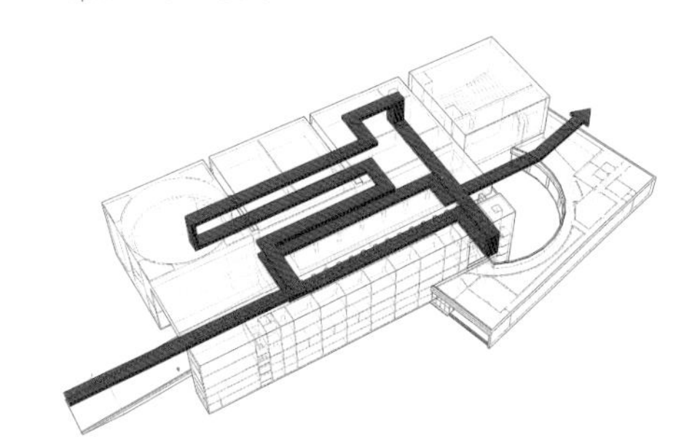

参观厅流线
Visiting Route

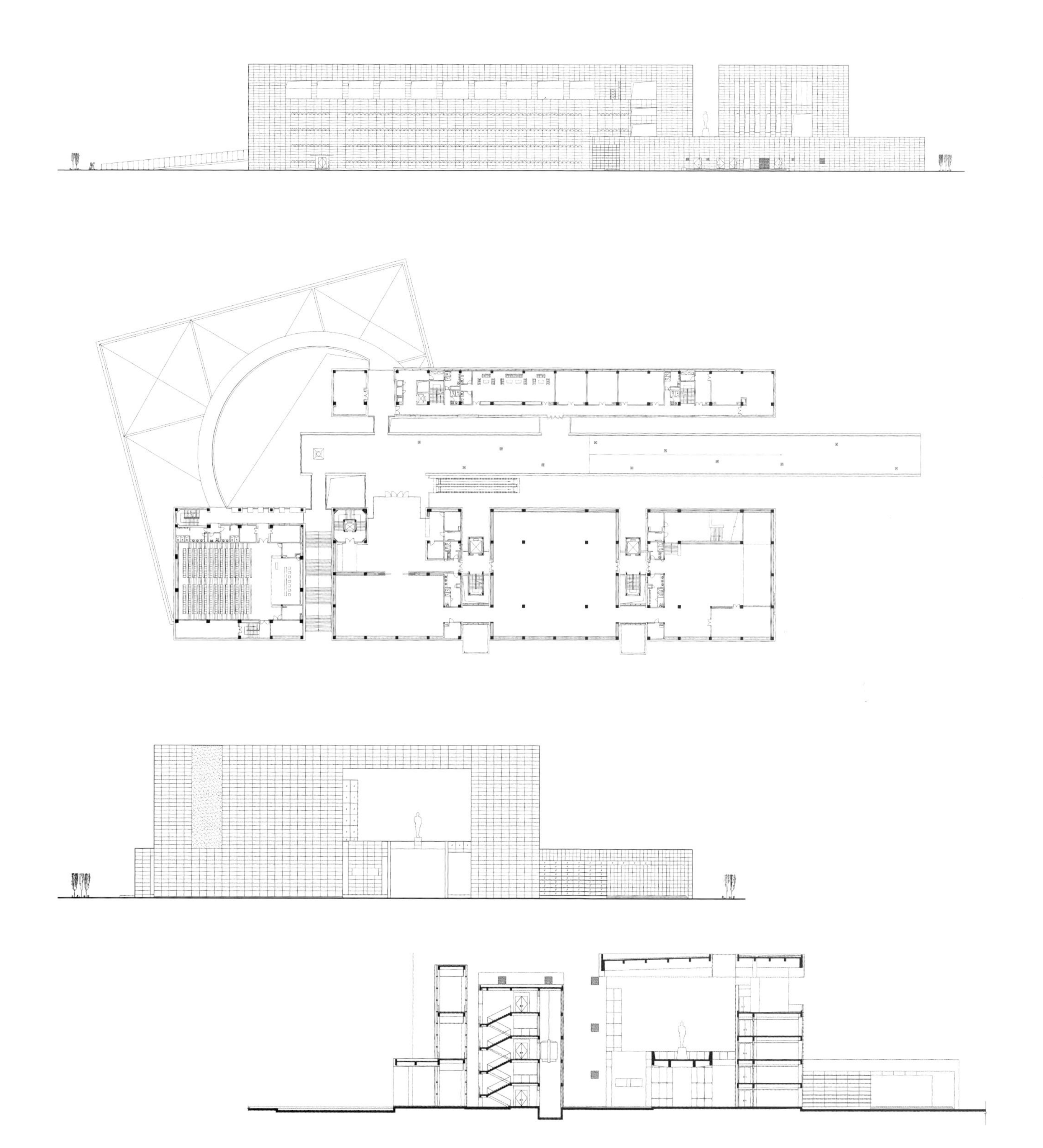

上至下：东立面图、二层平面图、南立面图、剖面图。
Top Down: East Elevation, Second Ground Floor, South Elevation, Section.

14 柳州工业博物馆
Liuzhou Industrial Museum

该项目位于第三棉纺织厂和苎麻厂的旧址，用地内建筑虽经时间洗礼，但保存较为完好，多为框架结构，砖砌外墙，极具时代感，布满爬藤植物的斑驳外墙向人们讲述着历史。

对原有建筑和柳州工业历史进行解读后，我们认识到，红色是柳州工业发展历史中的一个印记深刻的颜色，火红的钢铁、火红的天，象征着柳州人激情澎湃、敢为人先的工业精神。设计以红色作为主线，串联起一个个熟悉的场景，勾起柳州人民对那段火红的年代的回忆。

奖项：

广州市 2012 年度优秀工程设计一等奖；

广州市 2012 年度优秀工程设计风景园林三等奖。

This Project is built on the former site of the Third Cotton & Ramie Mill. Though weathering through many tough times, the buildings on the site is preserved in a good condition. The buildings here, most of them with a frame structure, reflect modern styles with its brick exterior wall which on the contrary relates the history with vines covering its face.

It is seen from the original buildings of the site and the industrial history of Liuzhou that the color of "red" is symbolic of the industrial development of the city: the fiery red steel and sky emblematizing the industrial spirit of Liuzhou — working as a pioneer with full passion! "Red" in the design threads the familiar scenes one after another to revive the memories of Liuzhou people of the past—the fiery red times.

Awards:

The First Prize for Excellent Engineering Design of Guangzhou, 2012.

The Third Prize for Excellent Engineering Design (Landscape) of Guangzhou, 2012.

总体鸟瞰图

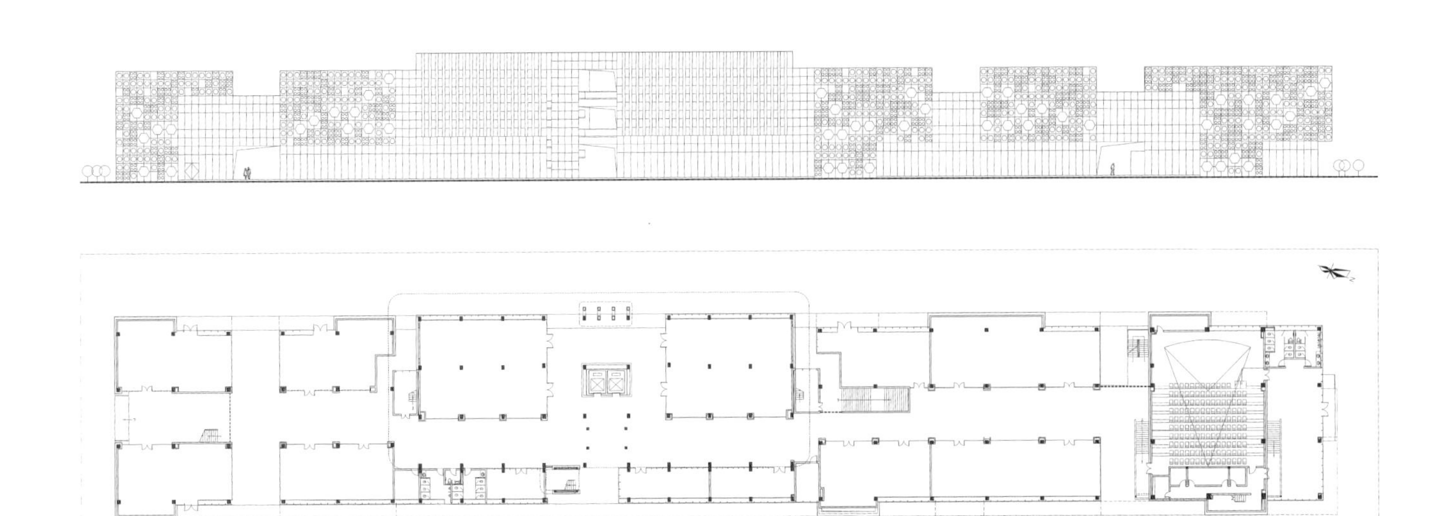

上至下：服务区立面图、平面图。
Top down: Elevation and Plan of Service Area.

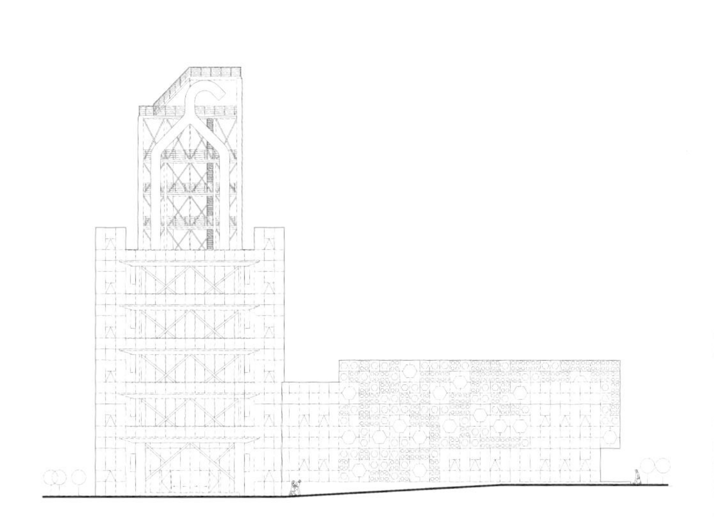

接待中心立面图
Elevation of the Reception Centre

柳州工业历史馆
LIUZHOU INDUSTRIAL HISTORY PAVILION

BLISS

把经济建设的立足
点转移到依靠科技
进步和提高劳动者
素质的轨道上

15 增城市中心医院
Zengcheng Central Hospital

本项目选址位于增城市经济技术开发区内，分为二期建设。一期包括门急诊、医技、住院综合楼，传染楼，职工饭堂。

设计首先是遵从城市设计原则和总体布局关系，以大气、简洁、统一的建筑语言，表达建筑群体的整体性和类型特征；其次是注重建筑的时代性和地域性的表达，以简洁的建筑形体、空间、材料诠释建筑的时代性，以空间渗透手法等诠释建筑的地域性，表达出岭南地区建筑的固有特色；再次是注重建筑设计的生态法则。方案借鉴传统岭南建筑的遮阳形式，汲取传统岭南建筑院落空间的通风、采光措施，从而实现建筑节能的功效，达到低碳、绿色的能耗目的。

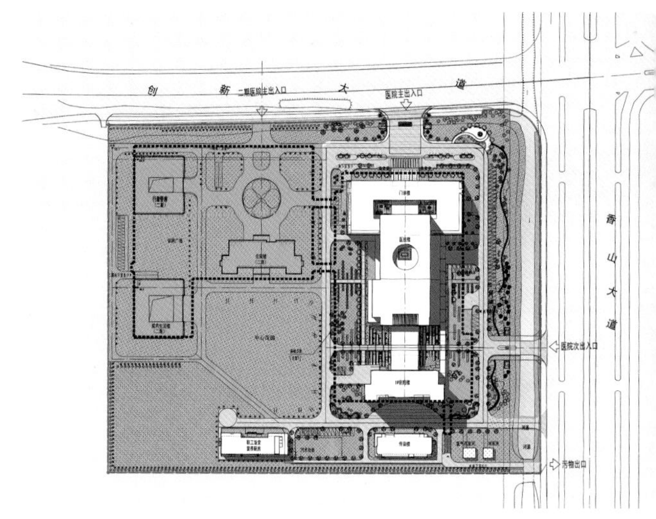

Located in Zengcheng Economic Development Zone, the Project is developed by two phases. Phase I involves Outpatient/Emergency Building, Medical Technology Building, Inpatient Building Complex, Infectious Disease Building and Staff Canteen.

Firstly, the design follows the urban design principle and general layout to present the holisticity and typical features of the building complex with generous, concise and unified architectural vocabulary. Secondly, it focuses on the building's feature of times and region, interpreting the former through concise building form, space and materials and the latter through courtyard and spatial penetration, thus showcase the inherent characteristics of buildings in Lingnan area. Thirdly, it emphasizes the ecological law of architectural design. With reference to the shading form of traditional buildings and the ventilation and daylighting measures of courtyard in Lingnan area, the design realizes the energy efficiency and the goal of low-carbon and green energy consumption.

16 广州市第八人民医院迁建项目一期工程
Phase I of Guangzhou No.8 People's Hospital Relocation Project

医院新址位于广州白云区嘉禾，距离广州市区约 15 公里。整个医院新址分三期进行建设，一期完成各项主要建筑，设置 200 张病床。

建筑设计突出了现代化、园林式，并具有岭南建筑风格。色彩运用和外形设计与周围环境相协调，妥善处理好功能、技术、形象的关系。通过外立面造型和内部空间处理及新型装饰材料的运用，使建筑物给人们以美观、清新的感受。

奖项：

广州市 2012 年度优秀工程设计三等奖。

Newly relocated in Jiahe, Baiyun District of Guangzhou with 15km away from Guangzhou downtown, the hospital is developed by three phases. Phase I involves various main buildings including 200 ward beds.

The architectural design highlights the modern features and garden-like layout with the architectural style in Lingnan area. The building color and appearance are harmonious with the surroundings and the design duly treats the relation among the functionality, technology and building image. The facade shape, interior spatial treatment and application of new decoration materials jointly contribute to the aesthetic and refreshing architectural effect.

Award:

The 3rd Prize for Excellent Engineering Design of Guangzhou, 2012.

市八医院

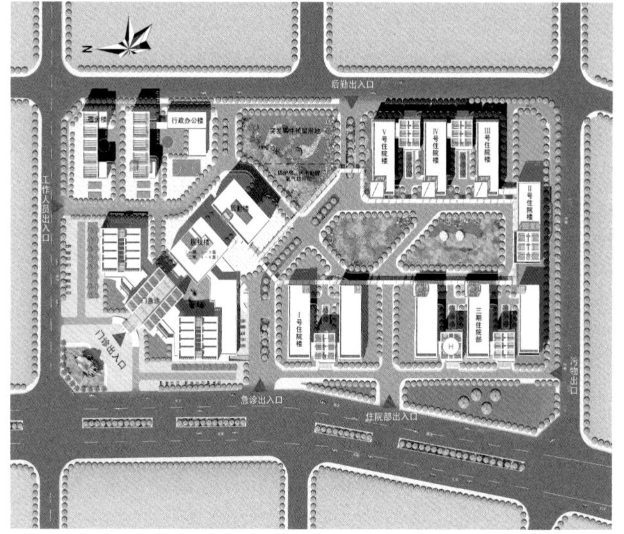
N
后勤出入口
行政办公楼
V号住院楼
IV号住院楼
III号住院楼
II号住院楼
I号住院楼
三期住院部
工作人员出入口
门诊出入口
急诊出入口
住院部出入口
污物出口

17 澳门大学珠海新校区
Hengqin Campus of the University of Macau (UM)

澳门大学新校区由3个用水网分隔的岛和主体学校、行政楼等书院式功能组团组成，功能组团包含12个学院，每个学院有自己的建筑群，包括课室、办公室、实验室、会议室和师生活动休憩空间。实行书院式管理；设有体育场、室内体育馆、室内泳池、3个开放式科研基地、音乐厅、行政楼及图书馆、书院组团及生活辅助设施等主要建筑。

Hengqin Campus of UM is composed of three islands, which are separated by the waterways and house the functional clusters including the campus and colleage-based functions like administrative buildings. The functional clusters accommodate 12 schools, each equipped its own building complex, including lecture theatres, offices, laboratories, conference rooms and lounges for staffs and students. Hengqin Campus implements a college-based management system. On the Campus there are such main buildings as stadium, indoor gymnasiums, indoor swimming

新校区强调“以人为本”，“可持续发展”，“现代化、信息化”，“园林式布局”，“糅合中西文化”的设计原则和理念。体现“中西荟萃，山海交融，岭南文脉，南欧风情”的建筑风格。

pools, three open scientific research bases, concert hall, administration buildings, libraries, college groups and daily supporting facilities, etc.

Hengqin Campus emphasizes the design phylosiphies of being people-oriented, sustainable, modernized/informationtized and features landscaped layout, and a mixture of Chinese and Western cultures, while presenting an architectural style that blends the Chinese elements with the western ones, the mountain with the sea, the traditional Lingnan traints with the southern European features.

18
邯郸汽贸城
Handan Auto Trading City

现代（邯郸）汽贸城项目由八个板块组成，每个板块自成体系，流通业务相互衬托，汽贸城整体规划和布局将汽车流通业态的全部功能融通，城区内道路、交通为汽车业态流通提供简捷通道和方便，现代（邯郸）汽贸城提供三个小时从购车到上牌（或检测、过户）"一站式"服务。

建筑是城市的一部分，是周边环境的延续，展销中心的设计从城市环境的界限和内容出发，对街道景观和城市绿化开放空间使用了两种建筑语言表达自我与整体的关系。建筑主体下部采用双层钢化夹胶 Low-E 玻璃幕墙设计，使展厅具有良好的采光面，减少对能源的依赖，建筑主体的上部使用白色 GRC 板幕墙，幕墙上开长条形的采光窗，简洁而富有现代感。

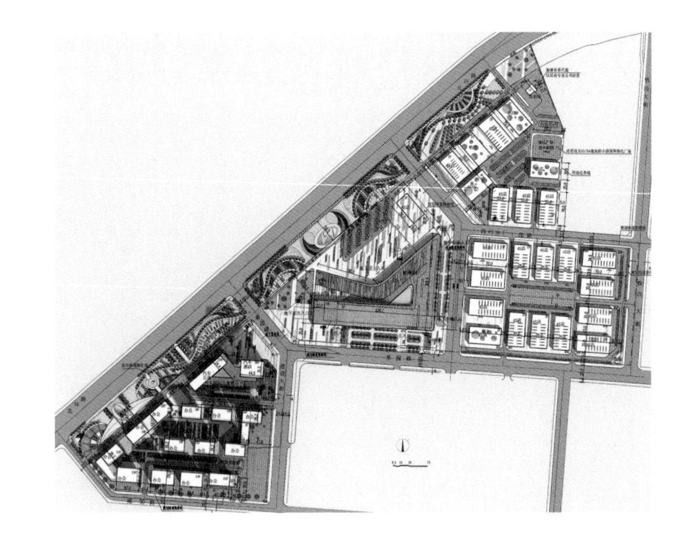

The Modern (Handan) Auto Trading City comprises eight blocks, each with self-contained business mechanism while complementing each other. The overall planning and layout of the Auto Trading City interconnect all the functions of the auto trading industry. The road and traffic inside the Auto Trading City provide the easy access and convenience for auto trading. The Modern (Handan) Auto Trading City offers the "one-stop" service from procurement to license registration (or examination and transfer) within 3 hours.

As the building represents the integral part of the city and extension of the surrounding environment, the design for the Exhibition and Sale Center, based on the boundary and contents of urban environment, employs two kinds of architectural vocabulary for the streetscape and urban green open space to express the individual and holistic relation. The lower part of the main building adopts the double layer tempered laminated low-E glass curtain wall to provide the favorable daylighting surface for the show room and reduce the reliance on the energy source. The upper part of the main building is clad with white GRC facade, on which the strip skylights are provided with concise and modern features.

会展中心 EXHIBITION CENTRE

ENGERMA

19 萝岗会议中心
Luogang Convention Center

萝岗会议中心（凯云楼）的建筑设计将景观元素渗透到建筑形体和建筑空间当中，以动态的建筑空间和形式、模糊边界的手法形成功能交织，探索使用功能之间的内在关系并使之有机相连，从而实现空间的持续变化和多样杂交。

建筑立面设计采取平实的设计手法，简洁的体块组合强化了现代会议中心的形象。同时采取架空等设计减轻了由会议厅功能所带来的较大的体量感，使建筑显得更轻盈。在建筑材料上采用石材为主要饰面，同时灵活地运用玻璃、钢材、可调遮阳百叶等进行搭配，力求使整个建筑群既传承了岭南建筑的精髓，又充满现代建筑的韵味。

奖项：

广州市 2010 年度优秀工程勘察设计二等奖；
广东省 2011 年度优秀工程勘察设计二等奖；
第十届中国土木工程詹天佑奖；
第二届广东省土木工程詹天佑故乡杯。

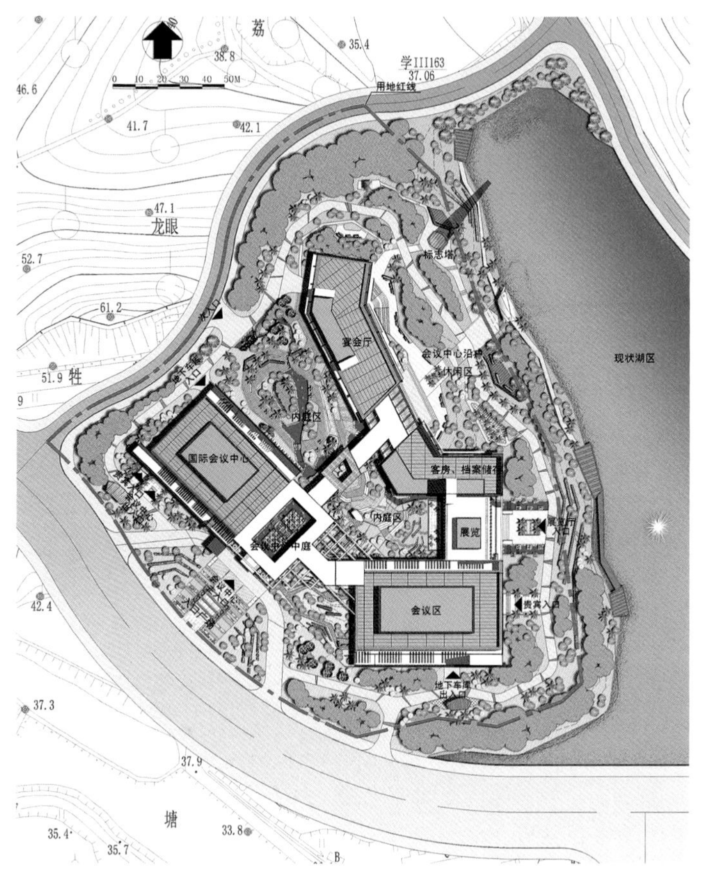

The architectural design of Luogang Convention Center (Kaiyun Building) integrates the landscaping elements into the building form and space. With the approach of dynamic building space/form and the blurred boundary, the design interweaves the functions, discovers and organically interlinks internal relations among the functions to realize the continuously changed, diversified and mixed space.

As for the building facade, the simple design and concise block combination enhance the image of modern convention center. The open-up design mitigates the gigantic building volume of the conference hall and realizes the light appearance. The stone, as the main finishes, flexibly mixed with the glass, steel and adjustable shading lamella makes the whole building complex both reference the essence of the buildings in Lingnan area and breathe the style of modern building.

Awards:

The 2nd Prize of Excellent Engineering Exploration and Design of Guangzhou, 2010.

The 2nd Prize of Excellent Engineering Exploration and Design of Guangdong Province, 2011.

The 10th Chinese Tien-Yow Jeme Civil Engineering Prize.

The 2nd Tien-Yow Jeme Hometown Cup for Civil Engineering of Guangdong Province.

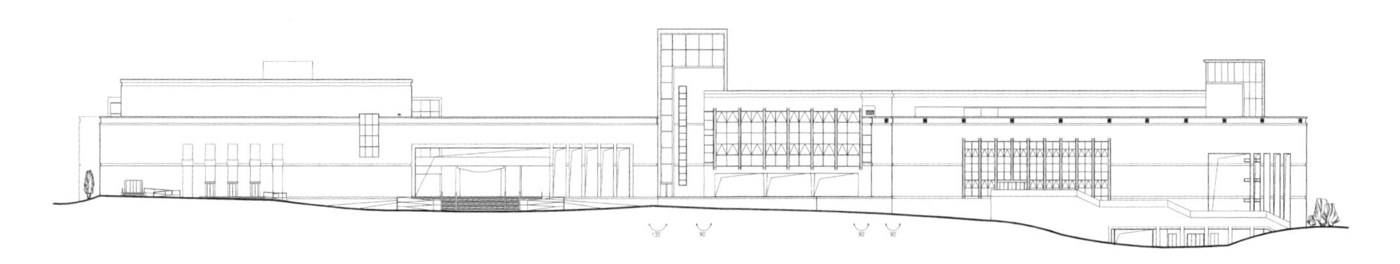

立面图、剖面图
Elevation and Section

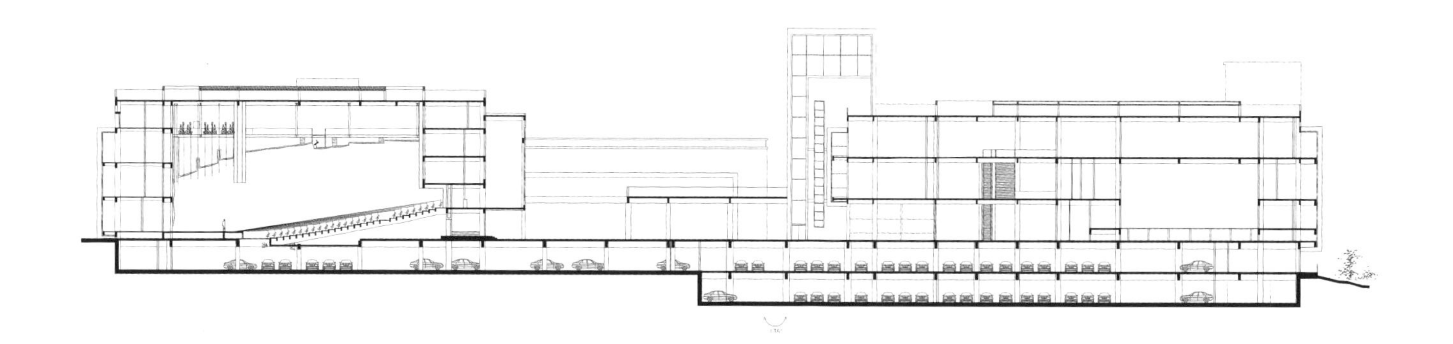

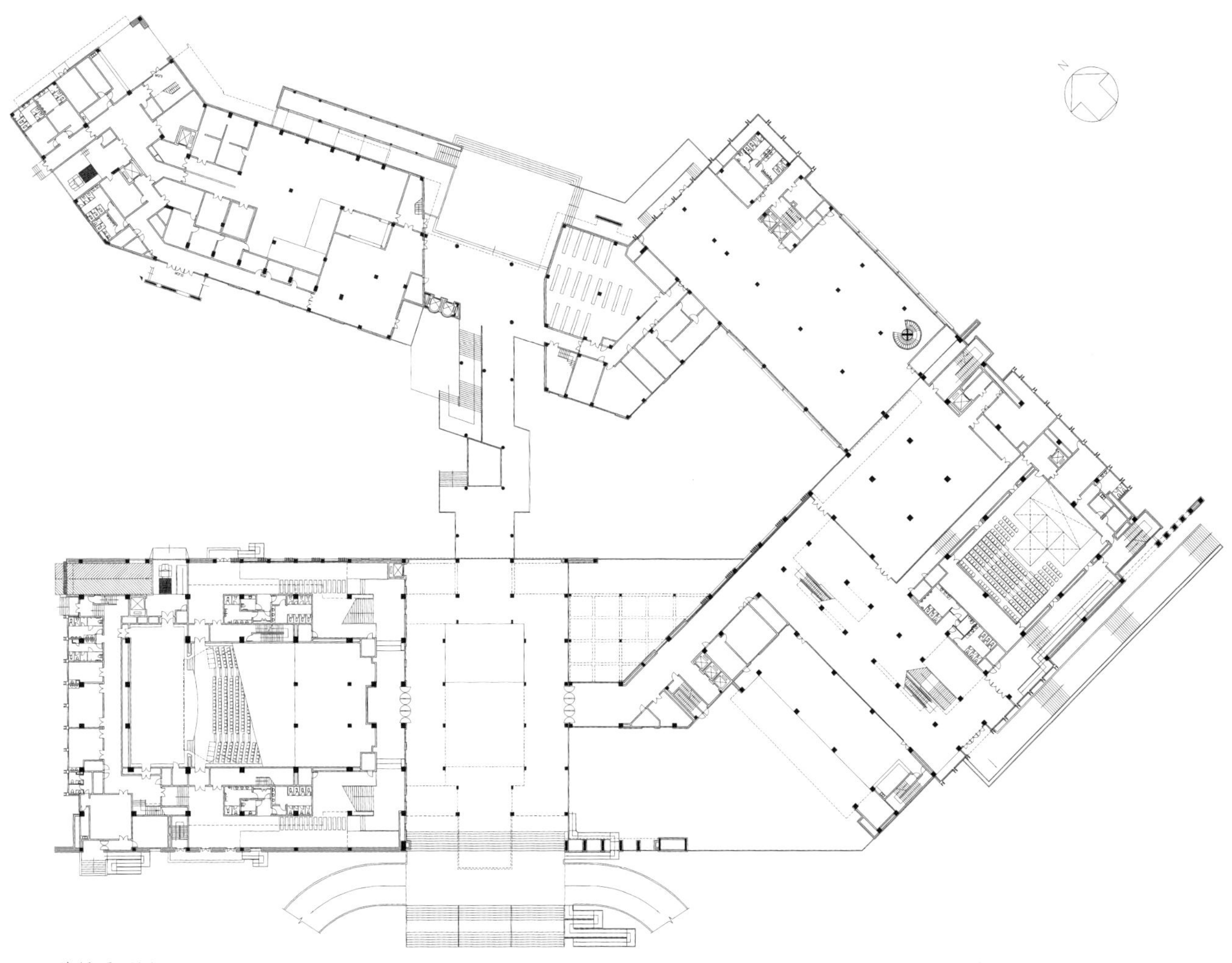

首层平面图
The First Floor Plane

20 广州白云国际会议中心
（与比利时 BuroII 建筑师事务所、中信华南建筑设计院合作设计）

Guangzhou Baiyun International Convention Center

(In collaboration with BURO II (Belgium) and Architecture Design Institute of CITIC South China)

广州白云国际会议中心是广东省广州市以会议为主体，辅以展览、商业配套用房等设施的大型综合性会议中心，是体现地方化特色，推动国际文化、经贸交流的重要场所。地上部分主要由 A、B、C、D、E5 栋建筑组成，我院主要负责承担会议中心 A、E 栋的方案深化、初步设计和施工图设计，D 栋的方案调整、施工图设计及地下室建筑专业的施工图设计。

奖项：

2008 年巴塞罗那世界建筑节公共建筑设计大奖；

2008 年国家优质工程银奖；

中国建筑业协会智能建筑优质工程；

第五届空间结构优质工程；

2008 年广东省优秀设计奖；

第八届中国土木工程詹天佑奖。

As a large comprehensive convention center, Guangzhou Baiyun International Convention Center mainly caters for the conference demands of Guangzhou city and Guangdong Province, meanwhile offers supporting facilities including retail and exhibitions. It is also an important place to present the local features and promote the international cultural and economic exchanges. The above-grade development mainly comprises Building A, B, C, D and E5. We are engaged to complete design detailing, DD and CD design for Building A and E, as well as design adjustment, CD design and basement architectural CD design for Building D.

Awards:

World Architecture Festival Award (Public Building), Barcelona, 2008.

Silver Prize of National Excellent Project, 2008.

Excellent Project (Functional Building) of Architecture Society of China (ASC).

The 5th Excellent Space and Structure Project.

Excellent Design Award of Guangdong Provide, 2008.

The 8th Chinese Tien-Yow Jeme Civil Engineering Prize.

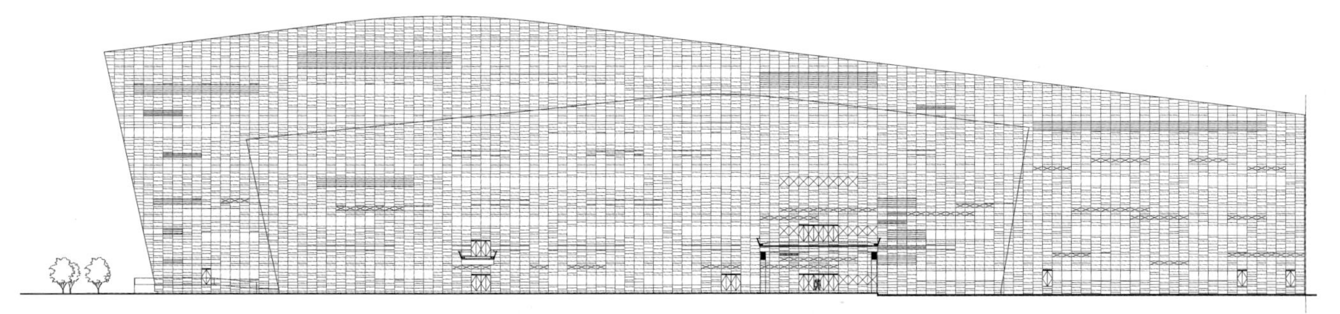

D栋立面图 D Elevation

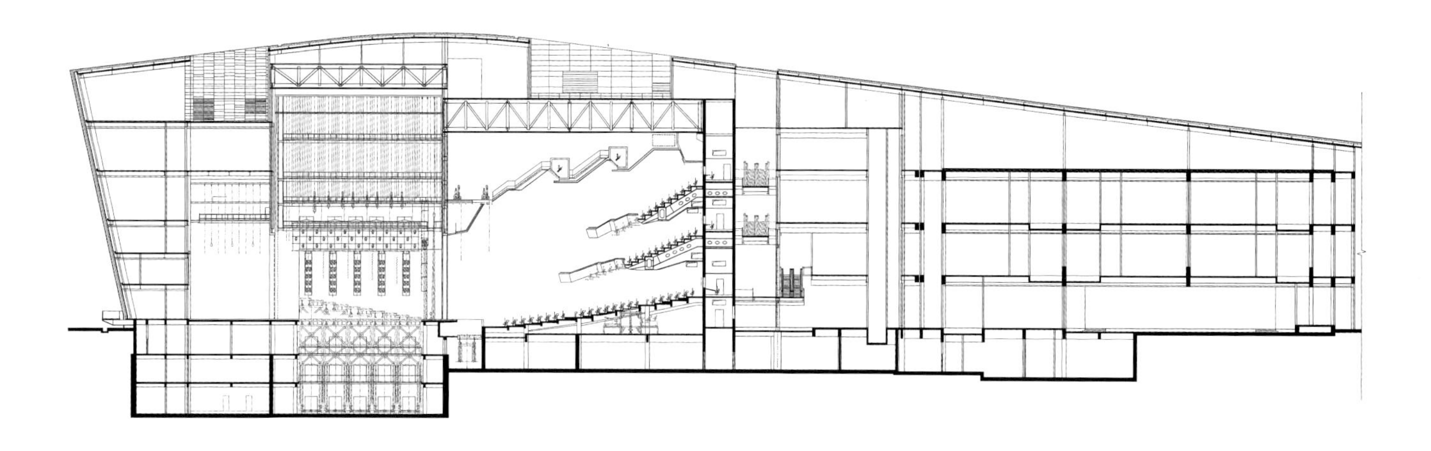

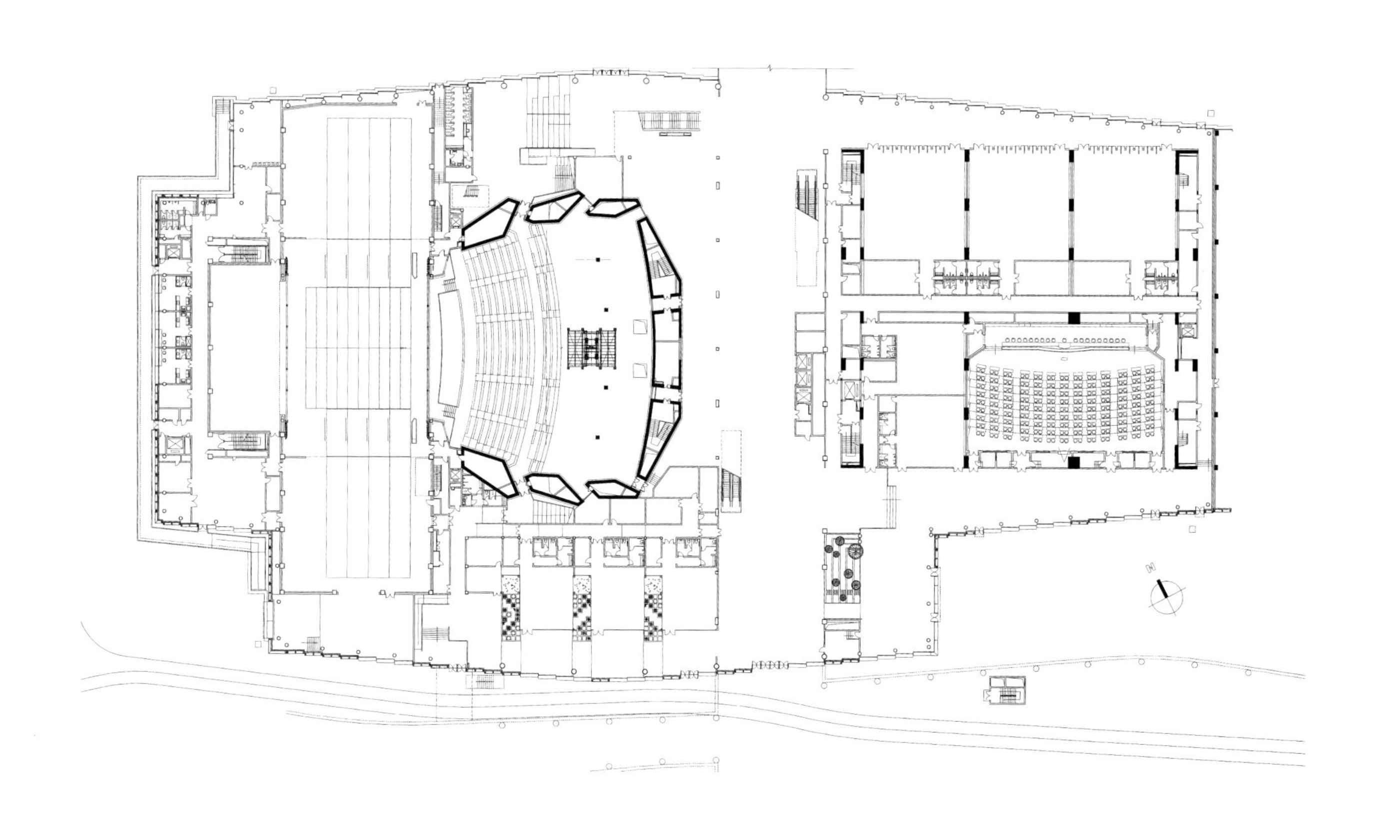

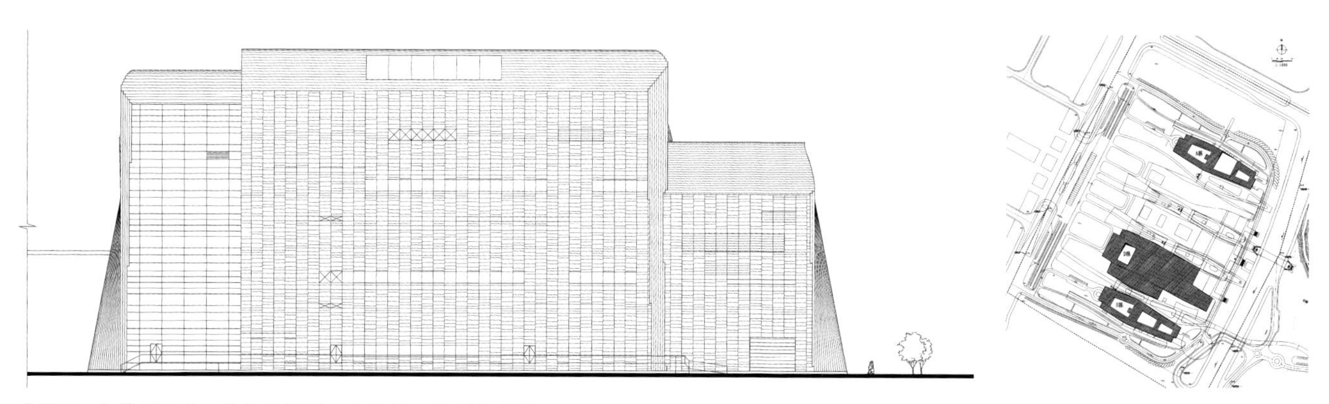

上至下：D栋剖面图、首层平面图、立面图、平面示意图。
Top Down: D Section, Ground Floor Plan, Elevation, Plan Stech.

21 广州市电视台新址
（与 GMP 国际设计公司合作设计）
Guangzhou New TV Station
(In collaboration with GMP)

广州市电视台新址位于广州市电视塔的西部，地理位置极为显要。鉴于电视塔建成后将成为世界第一高塔，因此，本项目设计应一方面突出电视塔的重要地位，而另一方面树立项目本身独立而鲜明的标志意义。

大楼主体的设计理念为两个长、宽、高均为 100 米的正方体体块，建筑体形清晰、简洁。立方体的建筑造型既包含了高度透明的空间，也有实体的建筑体量，二者相互交融，使整体造型看上去既稳重，又轻灵。

Located to the west of the Canton Tower (also known as Guangzhou TV Tower), Guangzhou New TV Station enjoys the superior geographic location. As the Canton Tower will rank among the tallest towers in the world, the prominence of the Tower should be emphasized on one hand; while on the other hand, the project itself has to stand unique and distinctive with its own iconic significance.

The main structure is designed into two 100m X 100m cubes with clear and concise building form. The cubic building features highly transparent space and solid building volume which integrate to each other and create the stable yet light building appearance.

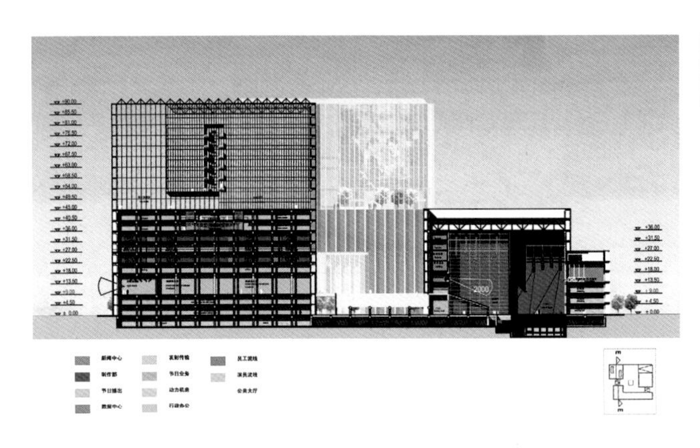

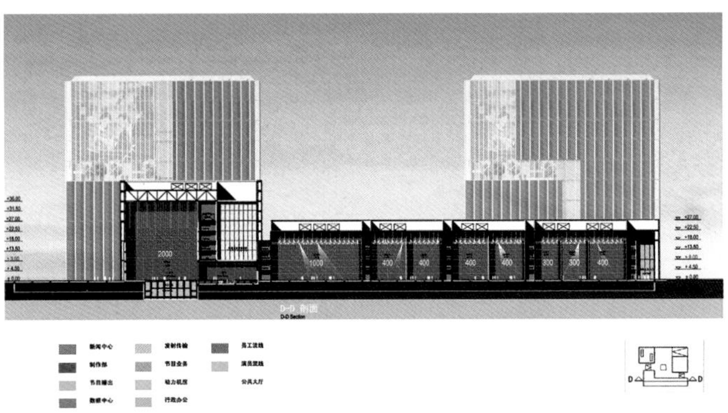

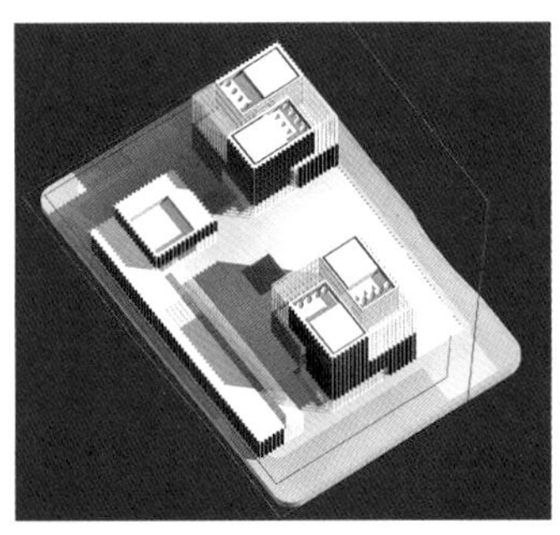

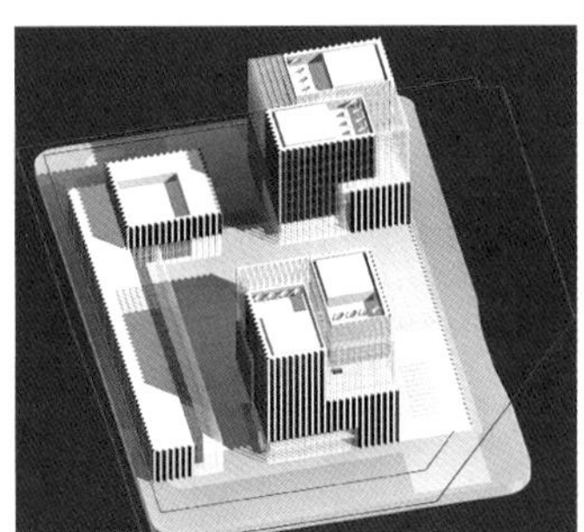

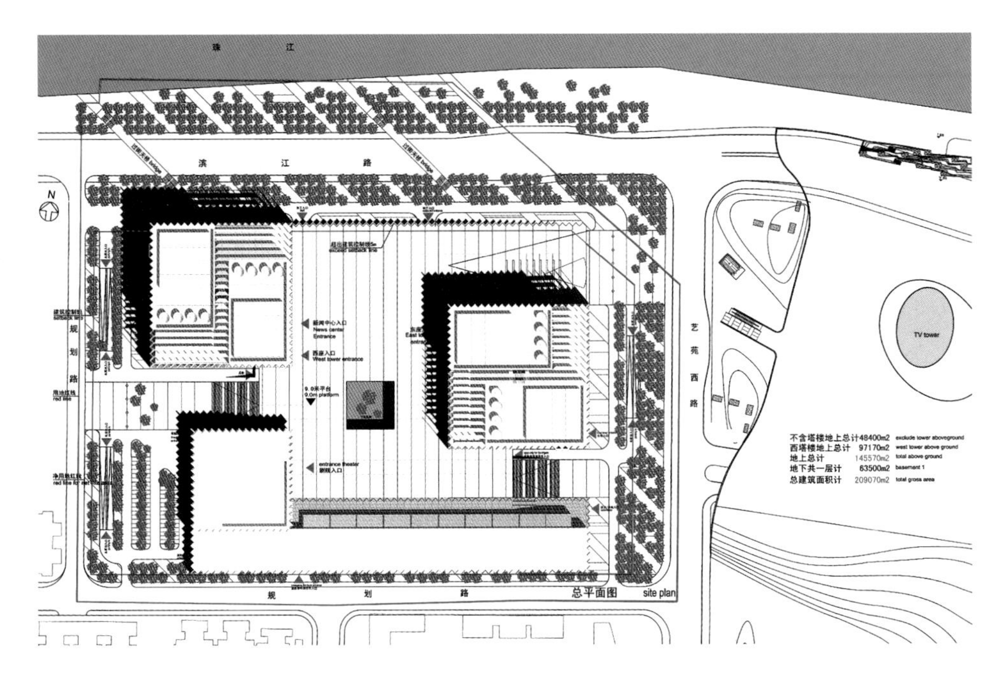

珠 江
滨 江 路
TV tower
艺 苑 西 路
规 划 路
总平面图 site plan
不含塔楼地上总计48400m2 exclude tower aboveground
西塔楼地上总计 97170m2 west tower above ground
地上总计 145570m2 total above ground
地下共一层计 63500m2 basement 1
总建筑面积计 209070m2 total gross area

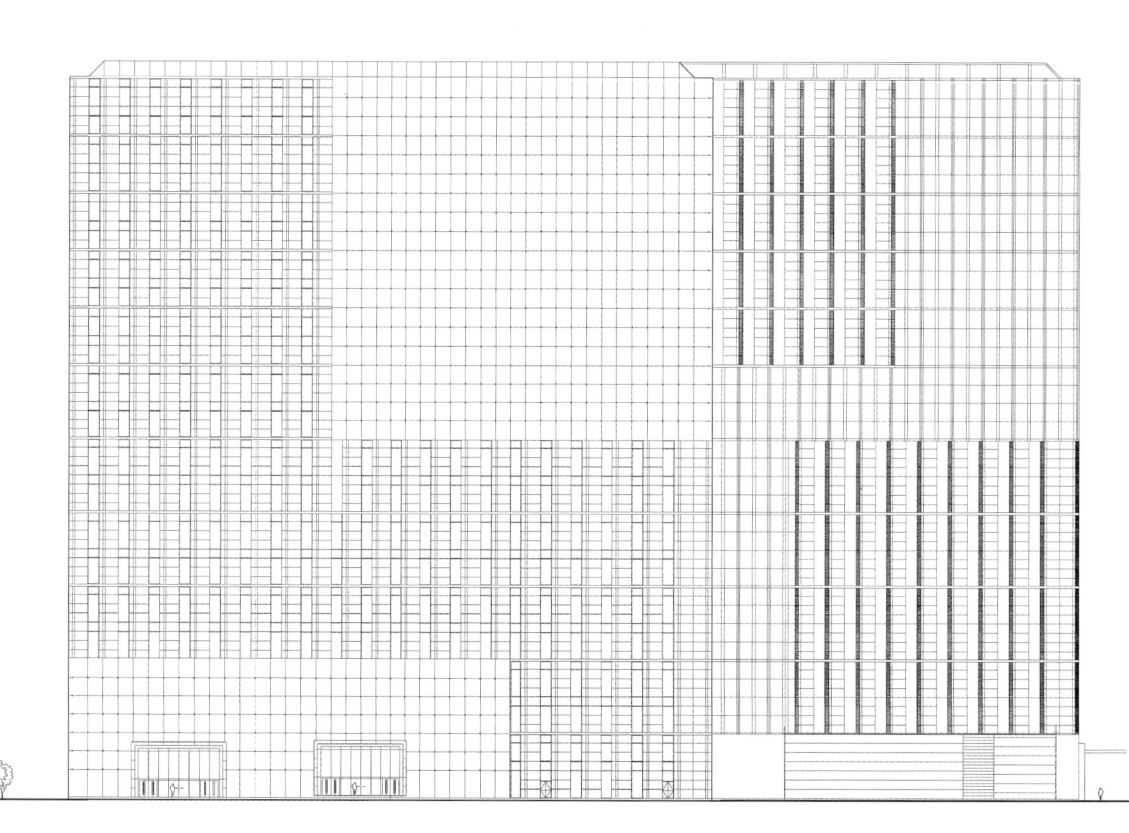

左下：东立面图
右上：首层平面图
右下：北立面图
Lower Left: East Elevation
Upper Right: Ground Floor Plan
Lower Right: North Elevation

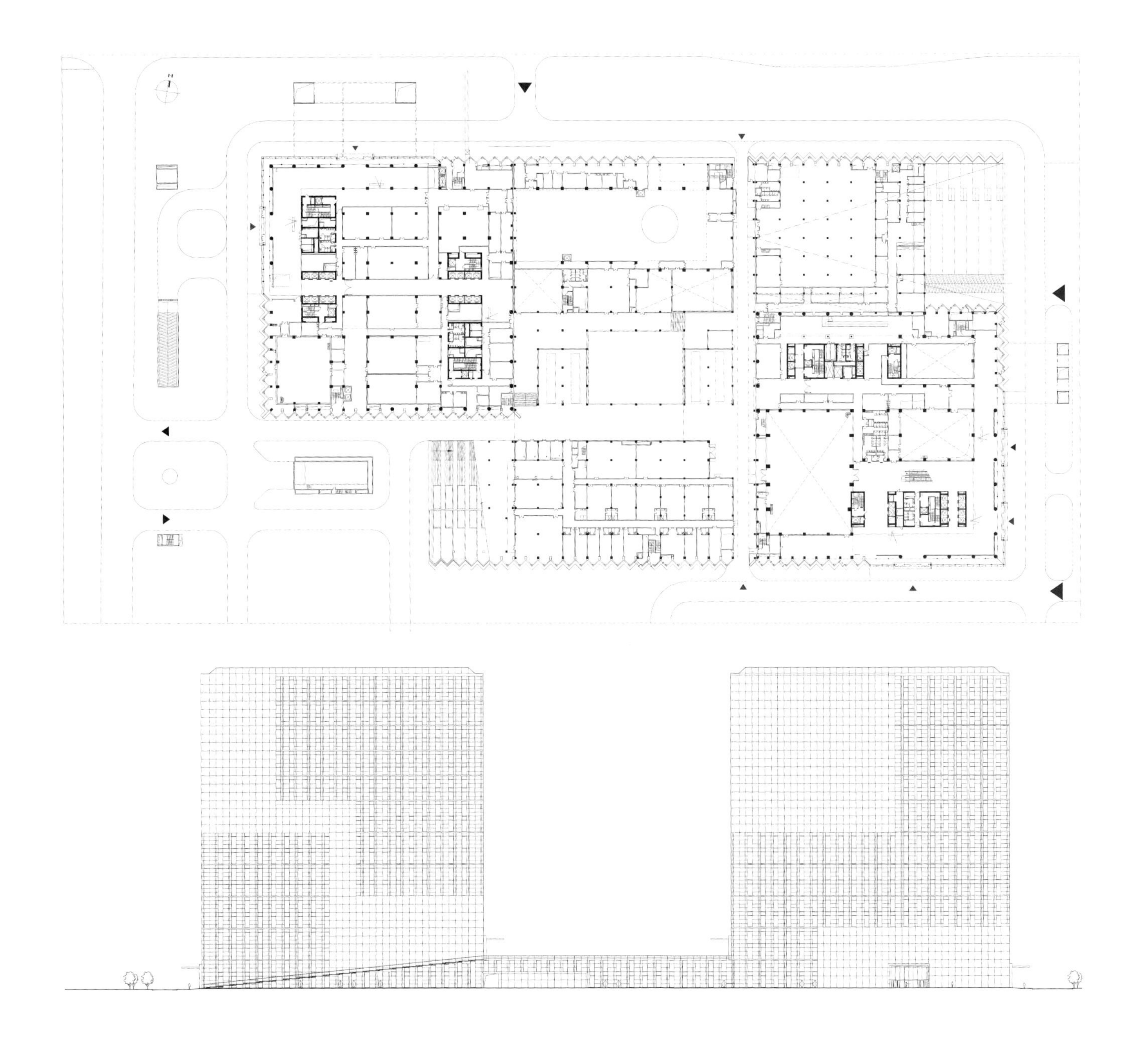

TRANSPORTATION AND SPORTS
交通与体育类

22

广州海珠客运站

Guangzhou Haizhu Passenger Station

广州被公认为千年不衰的东方古港，而海珠区四面环水，仿佛是漂浮在江上的航船。作为城市的门户，客运站具有重要的象征意义，从高空俯瞰，海珠客运站就像飘逸在珠江畔的一片绿叶，从侧面看又似一叶荡漾在碧波上的轻舟，有如展翼欲飞的“船形屋”，象征着广州历史悠久的“海文化”，延续了岭南建筑精巧通灵的地域文化生态。深远的挑檐板，高耸纤细的立柱，柔和优美的不锈钢装饰线条，大面积的通透点式玻璃幕墙及大跨度结构桁架的应用，使客运站具有“通、透、亮”的建筑风格。

奖项：

广州市 2004 年度优秀工程设计一等奖；

广东省第十二次优秀工程设计二等奖。

Guangzhou has been reputed as the oriental port enjoying a thousand year of prosperity, while Haizhu District in the city, embraced by water on all sides, resembles a boat floating on the river. A passenger station, as the gateway to a city, has important symbolic meanings. Overlooked from the sky, Haizhu Passenger Station can be likened to a green leaf dangling at the Pearl River, and to a boat on the river if viewed from the side. The Station, like a boat-shaped house on the wings, is emblematic of the time-honored "marine culture" of Guangzhou and carries on the local cultural ecology embodied by the exquisite Lingnan architecture. The station building is transparent, well-ventilated and light-flooded thanks to its expansive overhanging eaves, slender and upright columns, elegant stainless steel decorative stripes, extensive point-supported glass curtain walls and the long-span truss.

Awards:

The First Prize for Excellent Engineering Design of Guangzhou, 2004 .

The Second Prize for the 12th Excellent Engineering Design of Guangdong Province.

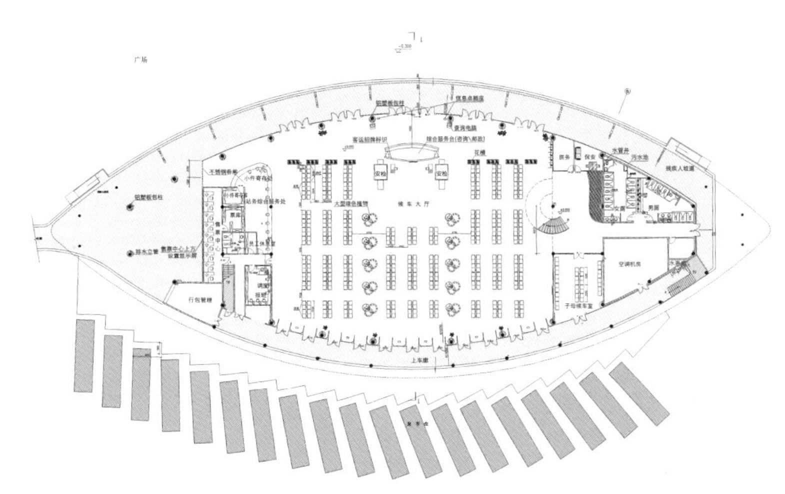

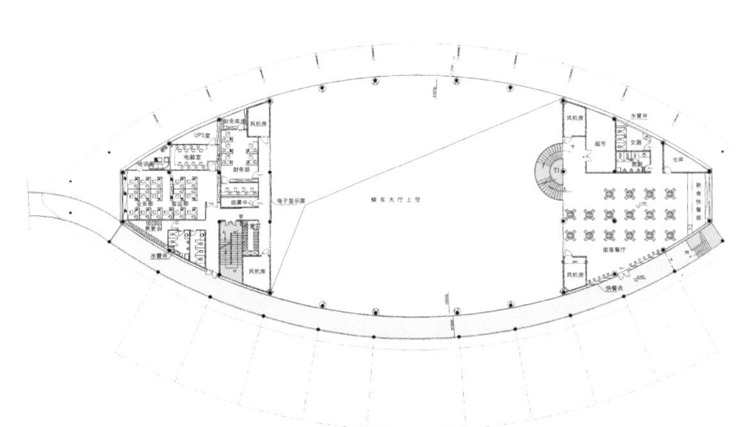

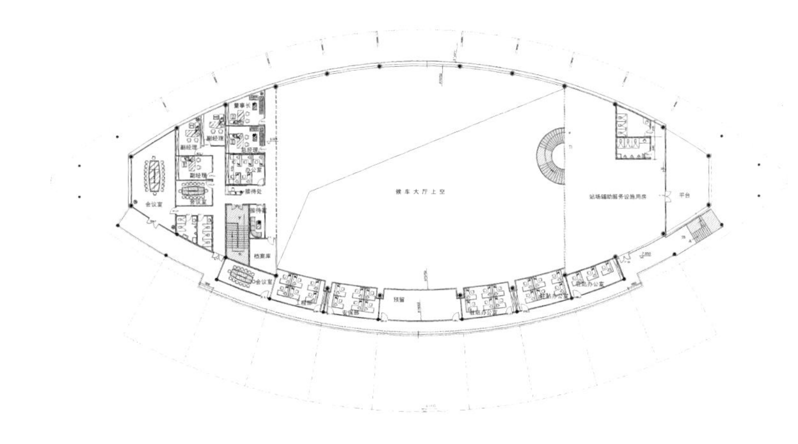

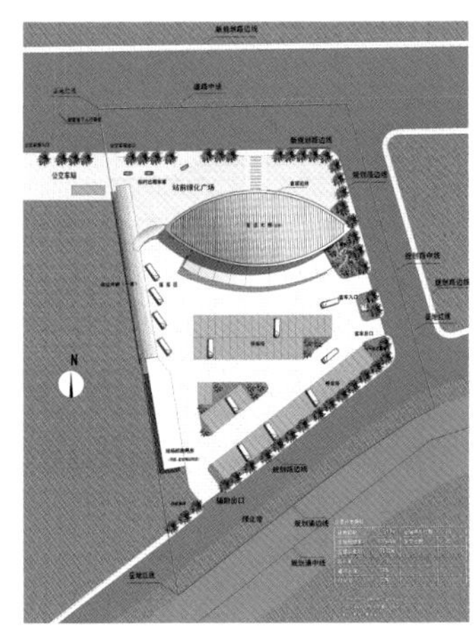

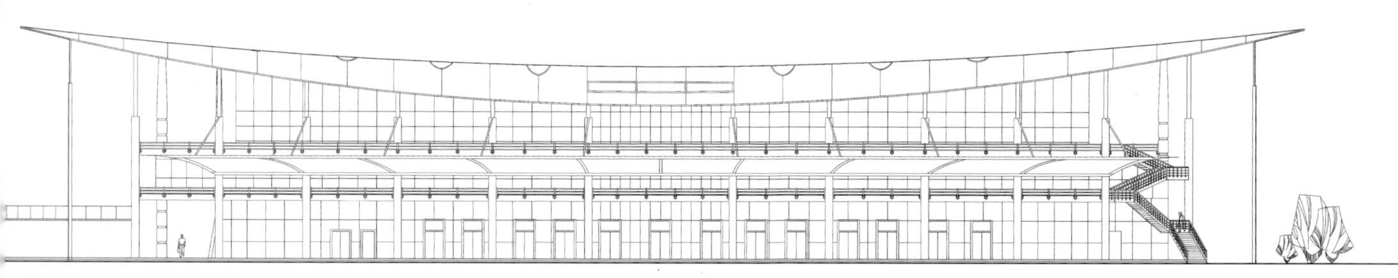

立面图 Elevation

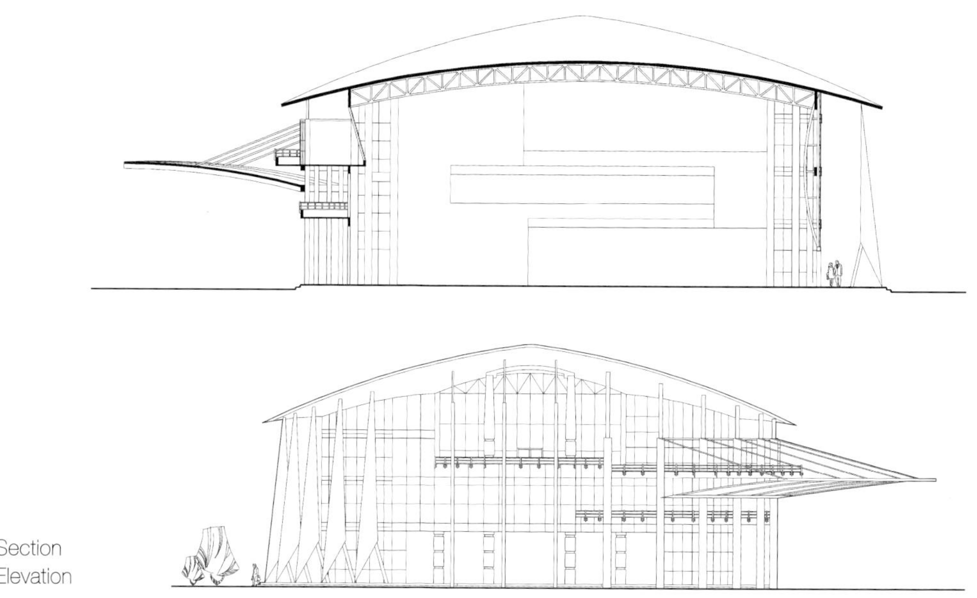

剖面图 Section
立面图 Elevation

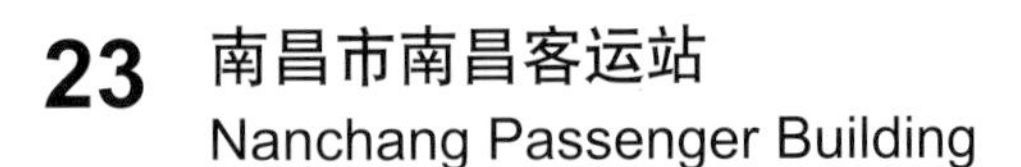

23 南昌市南昌客运站
Nanchang Passenger Building

南昌市南昌客运站是江西省政府为大力发展省会南昌市长中短途客运而建设的大型客运中心，同时也是南昌市地方标志性建筑。南昌客运站将成为南昌市客运枢纽，使长中短途客运和公交线合理驳接，满足公交与中长途大客车的停放、加油、维修需求。

客运大楼屋盖和外围护结构采用钢结构，内部建筑采用混凝土结构，柱网方整合理、间隔灵活，在力求节省造价的前提下又可使得内部功能布局实用、外部造型效果突出。建筑风格鲜明，富有个性，优美的建筑造型加上先进的建筑材料使各个建筑总体与周边环境和谐协调，展现出一系列合理的建筑空间效果和一个具有地标性的建筑群。

To vigorously develop long-middle-short distance passenger transportation, Nanchang government decided to build Nanchang Passenger Station, a mega passenger traffic centre and also a landmark building of Nanchang city. The project objective is to provide a passenger traffic hub for the city to facilitate the transfer between the long-middle-short distance passenger transportation and the middle-and-long bus lines while accommodating bus/coach parking, refueling and maintenance.

Steel structure is applied to the roof and envelope and concrete structure to the interior parts. The column grid is neatly and properly arranged to realize flexible subdivision, practical interior functional layout and a prominent appearance in a cost-effective manner. While presenting a uniquely individualized style, the buildings stay in harmony with the surroundings with the graceful form and up-to-date buildings materials, creating a series of rational building spaces and a landmark architectural complex.

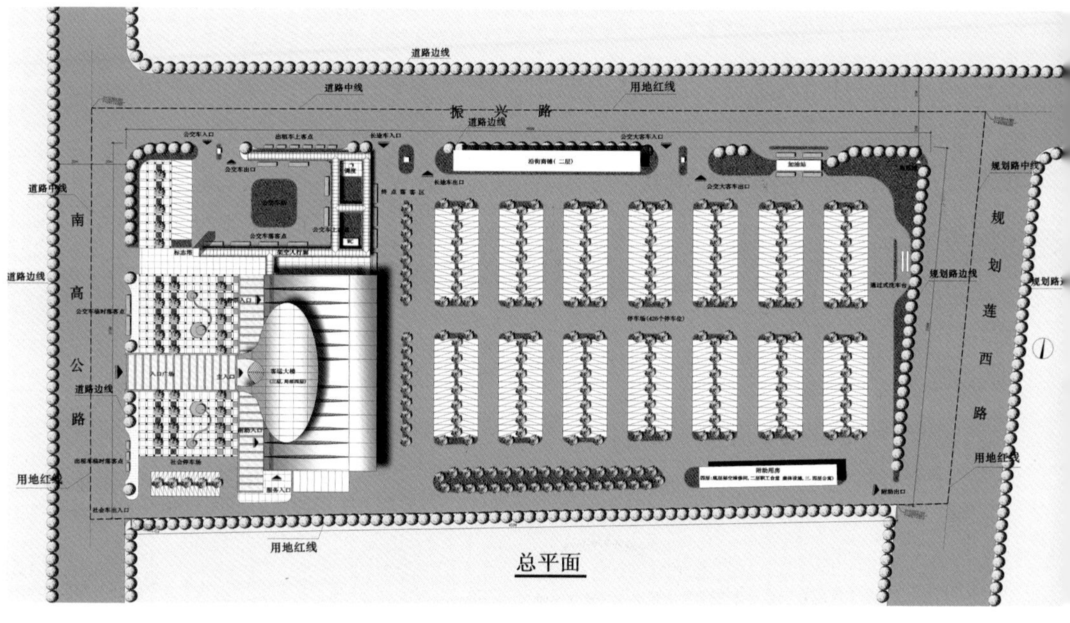
道路边线
道路中线
用地红线
振 兴 路
道路边线
公交车入口
出租车上客点
长途车入口
公交大客车入口
沿街商铺（二层）
加油站
公交车出口
长途车出口
公交大客车出口
规划路中线
道路中线
南
高
公
路
道路边线
终点落客区
公交车落客点
标志塔
架空人行廊
规划路边线
规
划
莲
西
路
停车场(426个停车位)
入口广场
主入口
客运大楼
附助入口
社会停车场
附助用房
服务入口
附助出口
用地红线
社会车出入口
用地红线
道路边线
用地红线
总平面

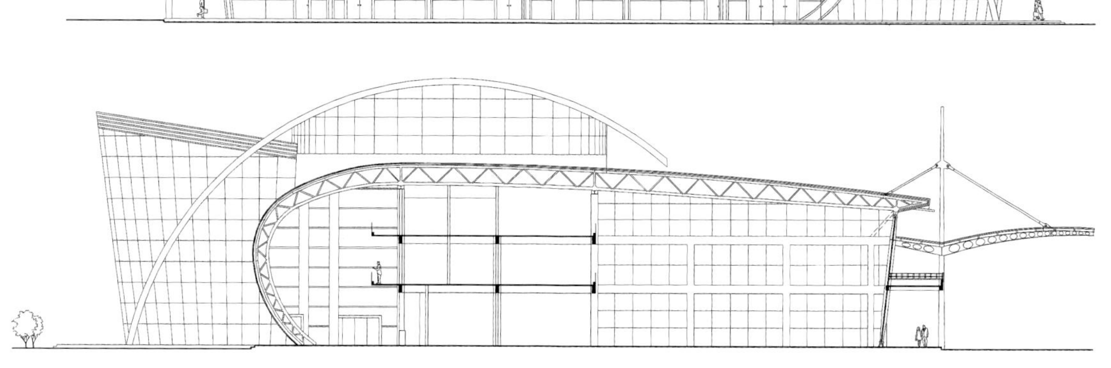

上至下：首层平面图、西立面图、北立面图、剖面图。
Top down: Ground Floor Plan, West Elevation, North Elevation, Section.

24 清远市城北汽车客运站
Qingyuan Northern Passenger Station

项目的主体建筑——客运大楼由前厅、候车厅、水上餐厅三部分组成，一个优美、流畅的弧线形屋盖覆盖的建筑沿清新大道徐徐展开，北高南低，充满动感，北面是一树叶形的通透玻璃体块与主体建筑取得平衡，树叶形的体块作为水上餐厅坐落在弧形的景观水池之上，周边环绕着水景、喷泉，成了广场上一个

The passenger terminal, main building of this Project, is composed of front concourse, waiting hall and on-water restaurant. The terminal, covered with an elegantly flowing arc roof, expands itself dynamically alongside Qingxin Boulevard with the north part higher and south lower. The on-water restaurant, a transparent leaf-shaped glass block on the north of the terminal to balance with the massing of the main building, appears to float on the arc reflective pool. Embraced with waterscape and fountains, the restaurant is indeed a highlight on

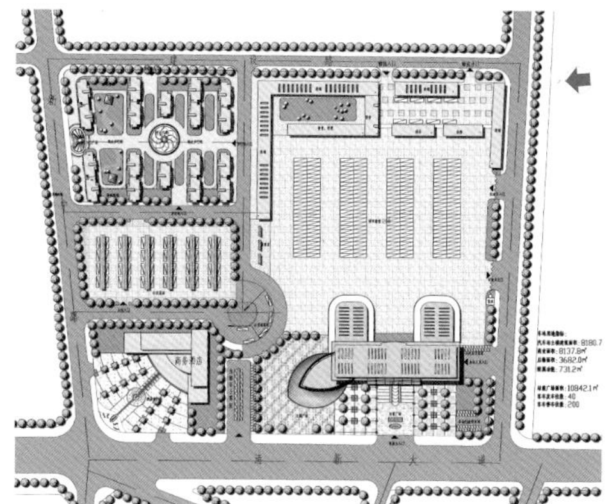

景观亮点，夜晚变幻的灯光和喷泉也为广场带来了另一番情趣。景观水池兼作消防水池。

奖项：

广州市 2008 年度优秀工程设计三等奖。

the square. The dazzlingly changeable night lighting and fountains bring exotic appeal to the square at night. The reflective pool functions concurrently as the fire prevention pool.

Award:

The Third Prize for Excellent Engineering Design of Guangzhou, 2008.

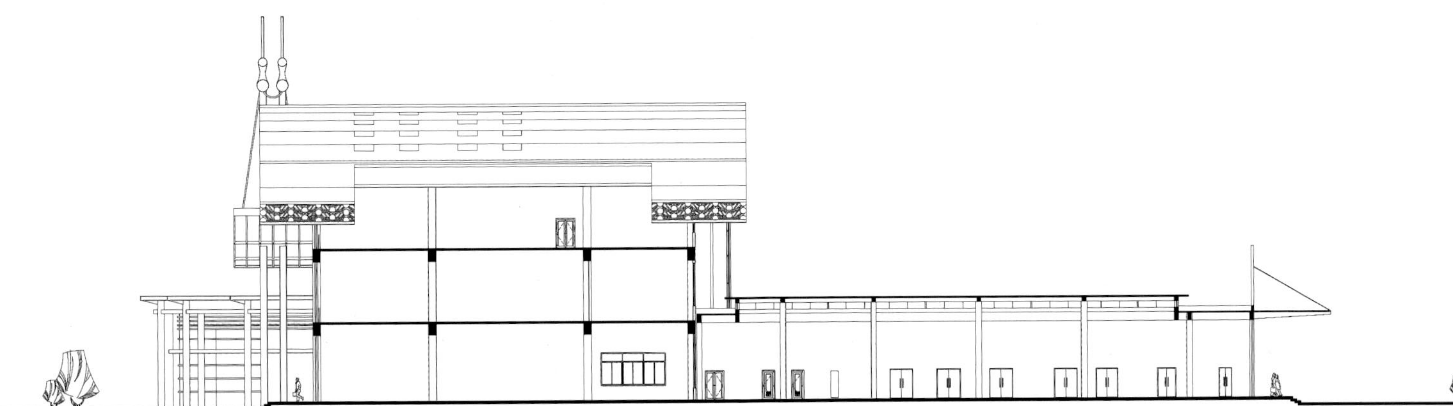

剖面图 Section

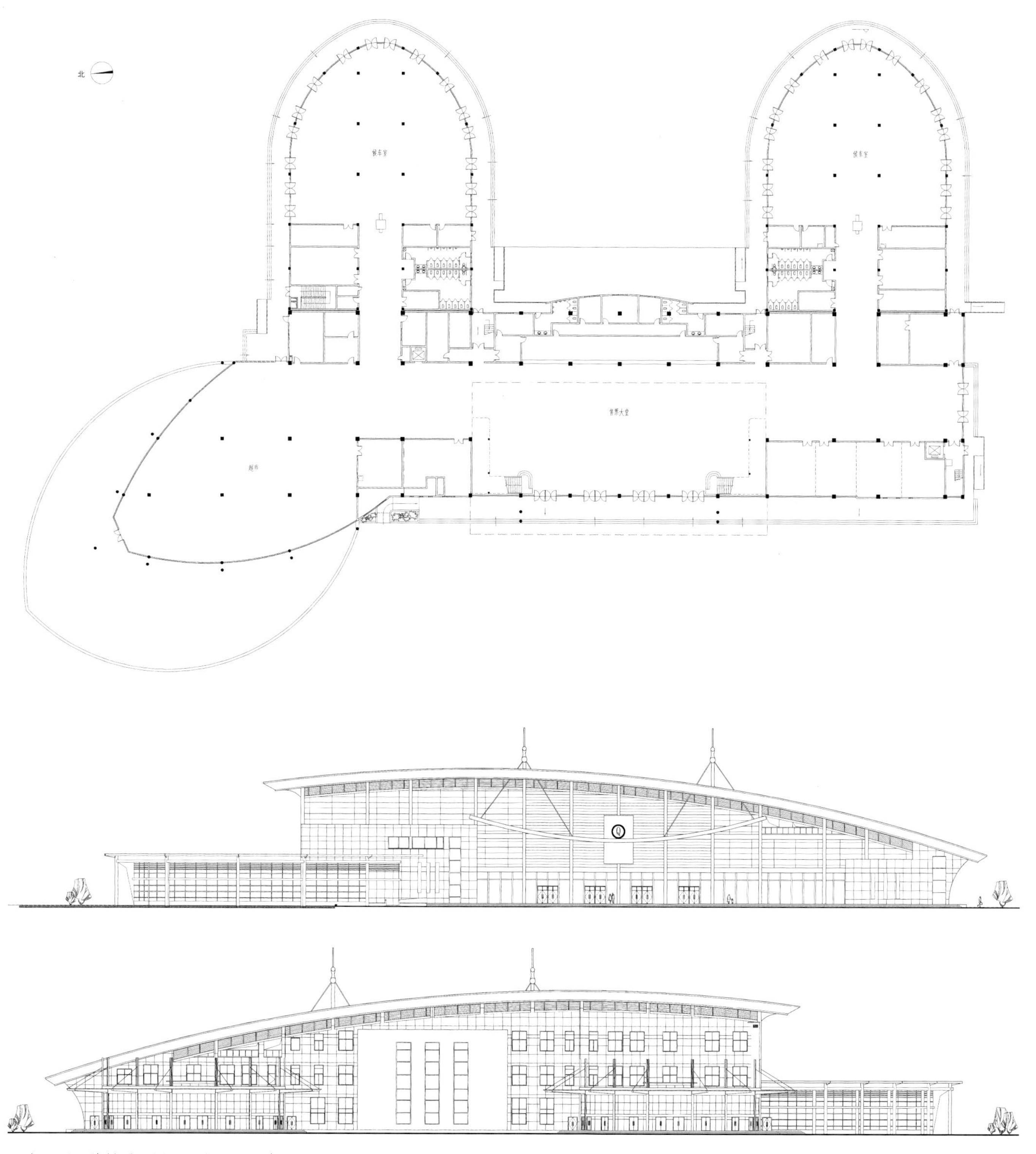

上至下：首层平面图、西立面图、东立面图。
Top down: Ground Floor Plan, West Elevation, East Elevation.

25 火车站房——苏州北站、徐州东站
Train Station Buildings–Suzhou North Railway Station & Xuzhou East Railway Station

列车在线路行驶的过程中，由于速度的不同，通过单位空间所需的时间是不一样的，而这种时空感，是乘客在列车里最强烈的一种感受。车站在这个过程中是一个特殊的节点，通过单位空间所使用的时间被非线性地放大了，这是一个时空爆炸的过程。苏州北站的设计，即是通过建筑形体的膨胀，对时空爆炸的呼应，同时在膨胀的过程中，在车站的中部形成了候车等使用空间。

When the train is running, it requires different time periods to pass the unit space due to different velocities. This enables the passengers on the train to have a strong time-space sense. The train station, during this course, is a particular singular point because the time required to pass the unit space here is non-linearly extended, namely the "big bang". Suzhou North Railway Station is designed as a swelling shape to respond to the concept of "big bang". The functional spaces, including waiting concourse, are created during this swelling process.

苏州北

徐州东

徐州东

26 澳门东亚运动会体育馆
Gymnasium for 2005 East Asian Games, Macau

此项目为澳门大型多功能体育馆，为 2005 年东亚运动会比赛、训练的主要场馆。场馆建筑面积为 6.8 万平方米，包括 8000 座的主体育馆、3000 座的小剧院及训练馆、5000 人大宴会厅（会展中心），造型新颖，科技含量高，纵向钢结构跨度为 380 米，横向达 280 米。

澳门东亚运动会体育馆设计立足于澳门地域文化的传承，用现代化的技术手段来表达澳门 2005 年东亚运动会的精神和内涵，使建筑与环境共生，反映澳门地区的地理和气候特征，充分展现时代的风貌。设计强调形体的完整性和协调性，力求以简洁大气的整体美感达到其可识别性，使之汇集建筑艺术与现代技术为一体，个性鲜明，在澳门城市形象建设中具有象征意义，成了澳门一道亮丽的城市建筑风景，一颗璀璨夺目的城市明珠。

奖项：

广州市 2006 年度优秀工程设计一等奖；

广东省 2007 年度优秀工程设计一等奖。

As a major multifunctional stadium for the competition and training of 2005 East Asian Games, the complex, with a GFA of 68,000 square meters, is composed of an 8,000-seat gym, a 3,000-seat theatre & training hall and a 5,000-seat banquet hall (conference and exhibition centre). This Project, with a vertical steel structure span of 380m and horizontal one of 280m, showcase the charm of high-tech with a novel appearance.

The gym design, rooted in the local culture of Macau, expresses the spirit and meaning of the Games with modern technologies and plays the modern charm to the fullest. The complex stays in harmony with the environment by reflecting the geographic and climatic features of Macau. In the design, the integrity and harmony of building forms are highlighted to realize the desired recognizability of the project through the concise and generous overall project image, and to integrate the architectural art and modern technology to establish a distinct identity for the project. The gym is another attractive view and icon that contribute to the city image of Macau.

Awards:

The First Prize for Excellent Engineering Design of Guangzhou, 2006.

The First Prize for Excellent Design of Guangdong Province, 2007.

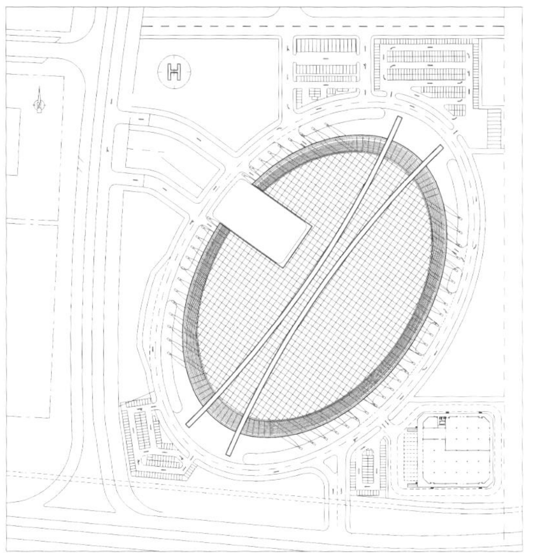

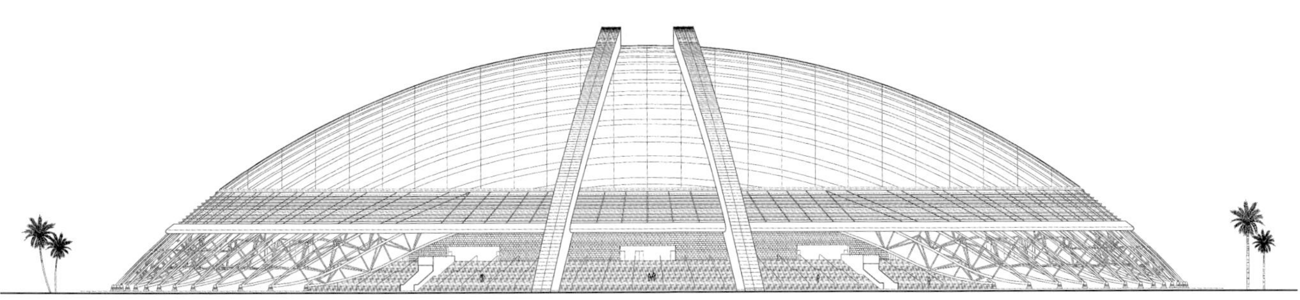

立面图 Elevation

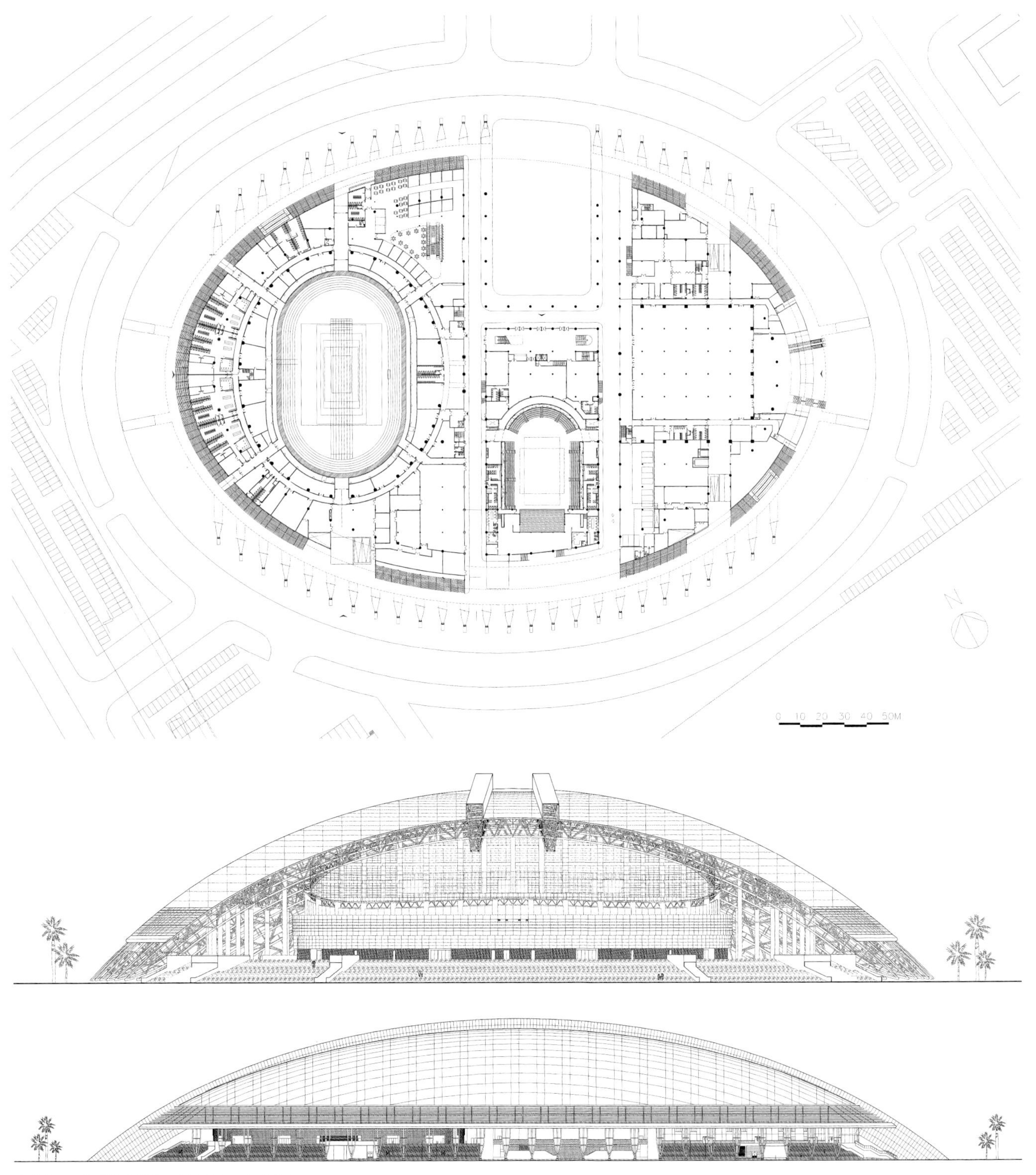

上至下：首层平面图、剖面图、立面图。
Top down: Ground Floor Plan, Section, Elevation.

27 澳门何东体育馆
Hedong Multifunctional Gymnasium, Macau

澳门是一个有着悠久历史文化又充满希望和朝气的地区，该体育馆的建筑是中西文化交流和古今建筑融汇的结晶，既有欧洲古建筑文化风韵，又有新时代、新技术的内涵。

体育馆的设计运用了传统的柱式排列和现代化的“V”形屋面构图手法，以新颖的建筑技术和材料营造出典雅、壮观、清新和亮丽的新澳门建筑风格。体育馆的建筑风格融合在现有的建筑环境中，又为繁荣的传统街区增色添彩。

奖项：

广州市 2006 年度优秀工程设计二等奖；

广东省 2007 年度优秀工程设计三等奖。

Macau is a time-honored historic city full of hope and vitality. The Gymnasium is an integration of Chinese and Western cultures as well as of ancient and modern architecture. This Project blends the European charm with modern Macau appeal and ancient architectural culture with modern technologies.

The Gymnasium incorporates the traditional colonnade with the modern V-shaped roofing, and, with the new building technology and materials, presents a neo-Macau architectural style of both magnificent elegance and fresh brightness.

Awards:

The Second Prize for Excellent Engineering Design of Guangzhou, 2006.

The Third Prize for Excellent Design of Guangdong Province, 2007.

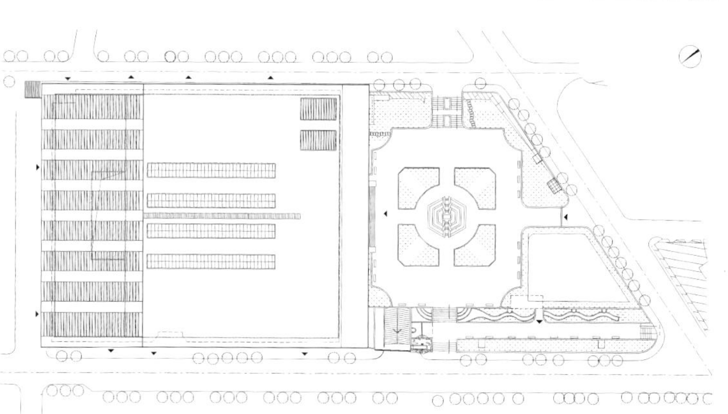

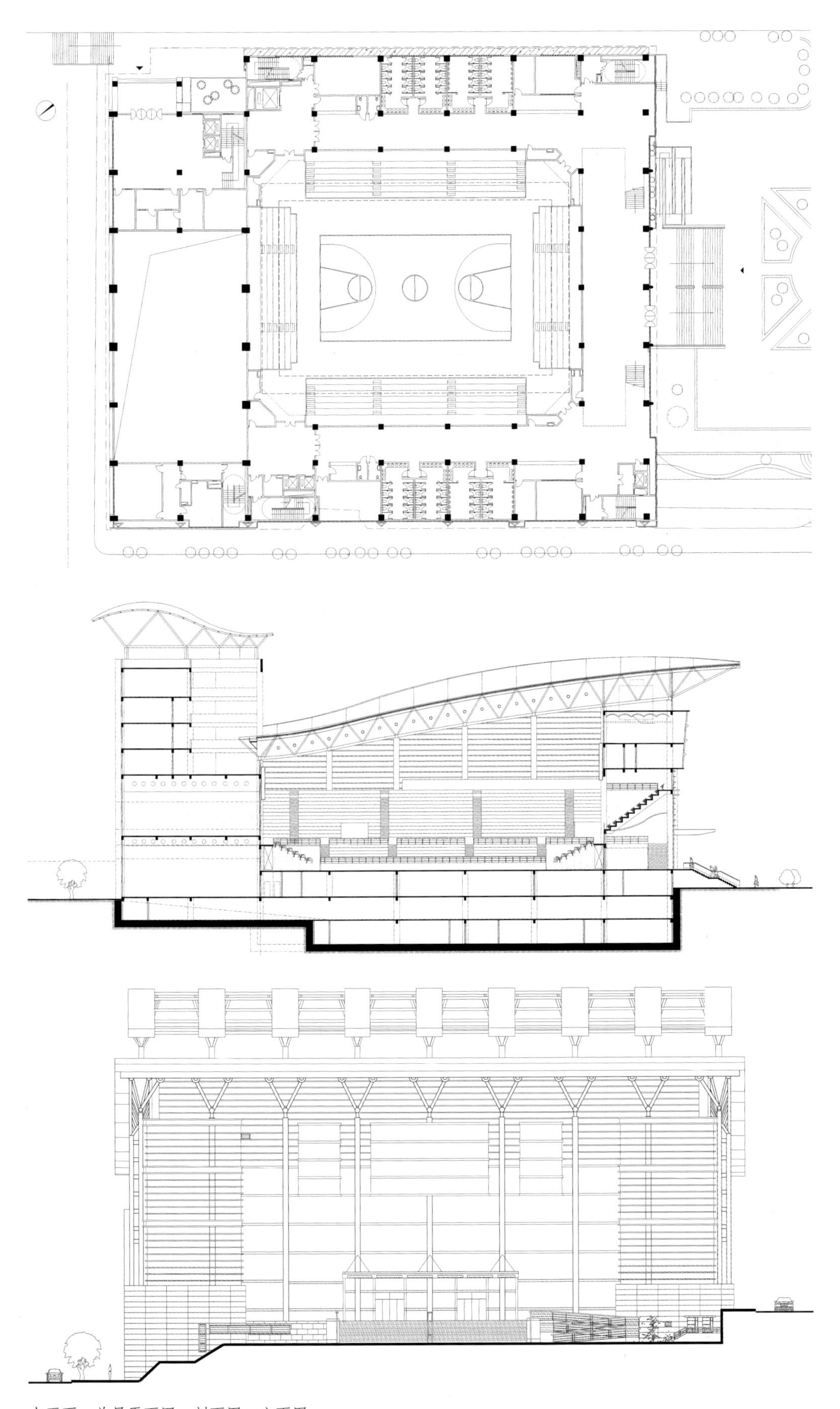

上至下：首层平面图、剖面图、立面图。
Top down: Ground Floor Plan, Section, Elevation.

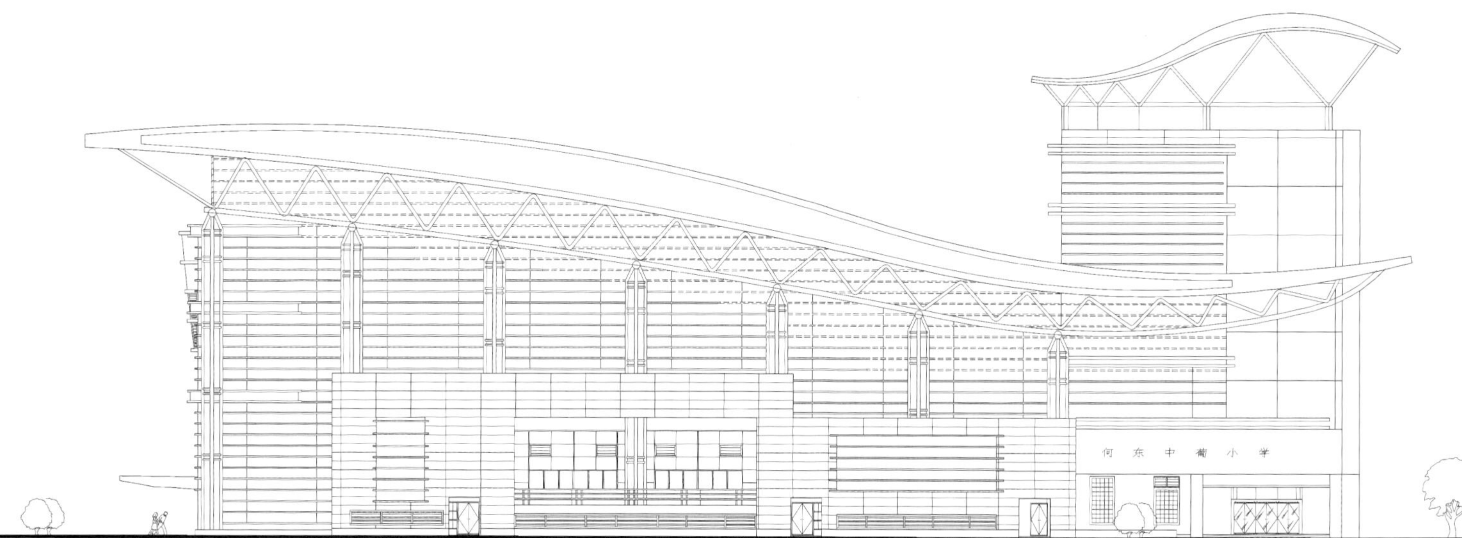

立面图 Elevation

28 广州亚运板球场
The Cricket Court of 2010 Guangzhou Asian Games

设计构思从板球场的特点——“圆”出发，体块以圆环状紧抱场地，随功能的嵌入、两端的抬高，以 90 度截面为轴进行扭转，形成蜿蜒起伏的独特建筑形态。

以曲线为母题，一组宛若浮云的建筑群形体简洁而富有象征性，给人以强烈的动感及视觉震撼。建筑立面肌理取义于“星旋”，速度条的设计使建筑充满了速度感，以“动态瞬间”表意“更快、更高、更强”的体育精神。通过绿色廊道及绿林公园等生态设施建设，在为板球场观众提供良好绿色环境的同时，通过可固碳释氧的绿林，共同打造亚运生态绿色场馆。

奖项：

广州市 2012 年度优秀工程设计二等奖。

The design is initiated from the round shape of the Cricket Court. A ring-shaped building block closely embraces the Court. With the embedded functions, the both ends of the buildings are raised and then twisted around the axis of 90° section, presenting an undulating building form.

With curvilinear lines as the motif, a series of building blocks, concise and symbolic, resemble floating clouds and bring strong visual impact and dynamics. The facade fabric is inspired by the "rotating stars". The "speed stripes" are designed to inject sense of speed into the building and express the spirit of sport, i.e. "higher, faster and stronger" with the "dynamic moment". The ecological facilities like Green Corridor and Green Wood Park provide the spectators with refreshing environment. The green woods, with carbon sinking and oxygen emitting functions, contribute to an ecologically friendly Cricket Court for the Asian Games.

Award:

The Second Prize for Excellent Engineering Design of Guangzhou, 2012.

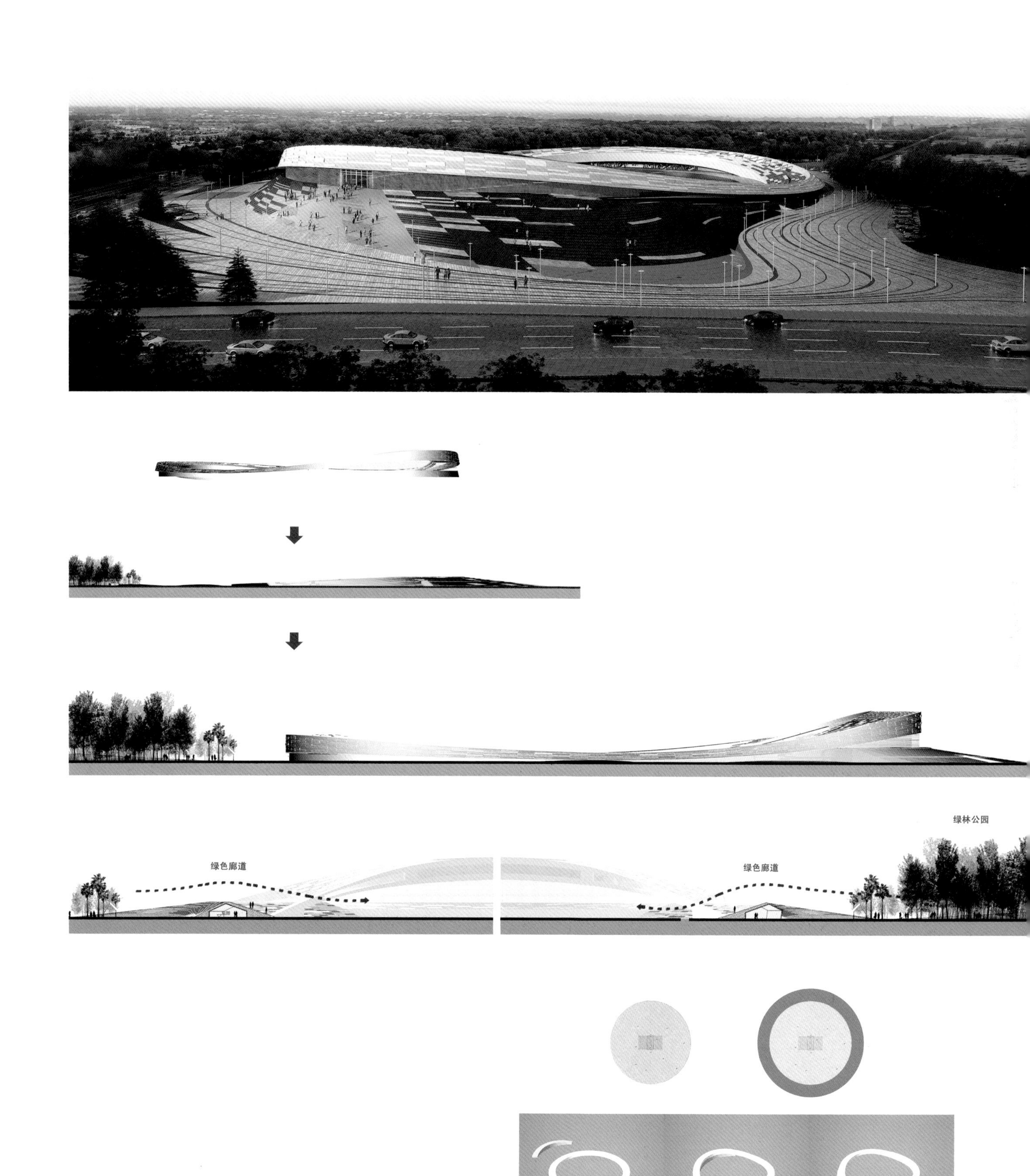
绿林公园
绿色廊道
绿色廊道

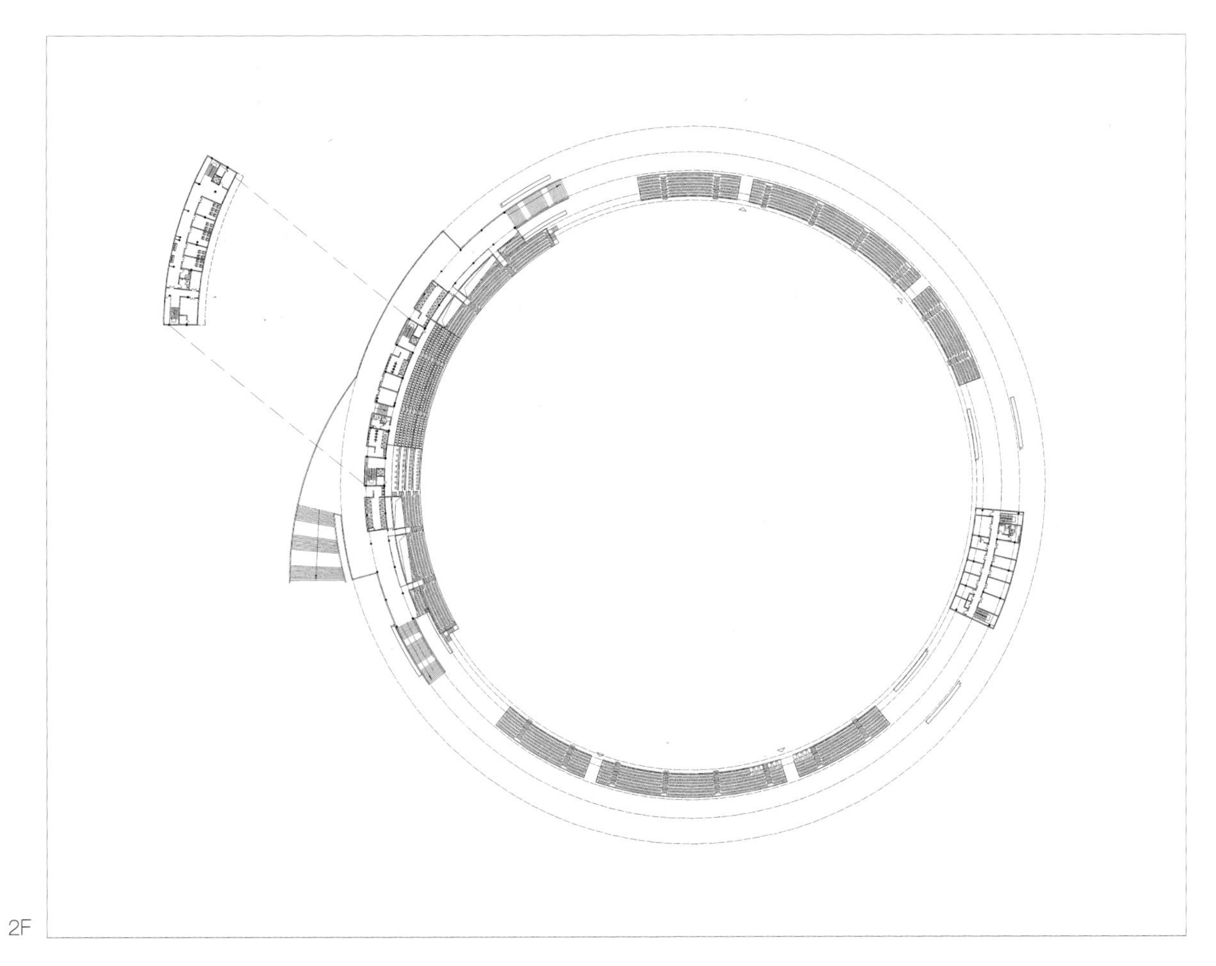

2F

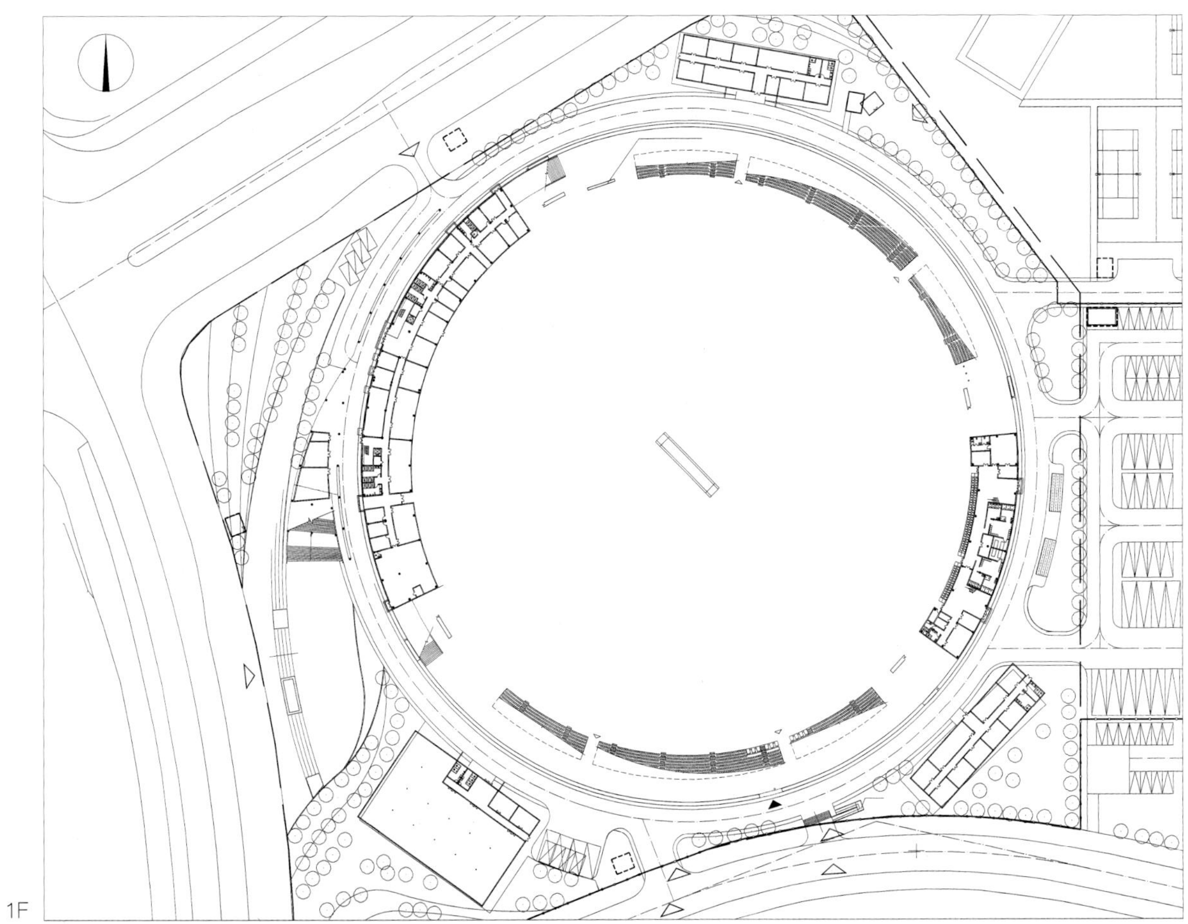

1F

顶视图 Top View

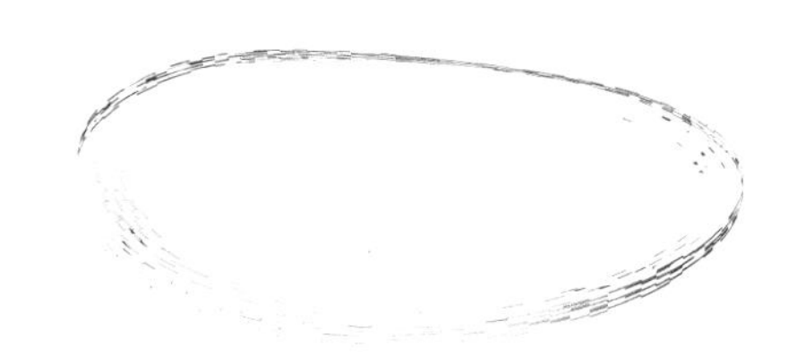

侧视图 Site View

29 广州棋院
Guangzhou Chess Institute

本项目位于广州市白云山风景区的横枝岗路，广州市麓湖公园东侧，西、北面为白云山，东邻广空军事用地，南面为白云山权属地块。

以“院静春深昼掩扉，竹间闲看客争棋”的意境为主题，采用棋盘格局为基本规划设计母题，将建筑化整为零，以灵活多变的墙体元素贯穿其中，而建筑功能与庭院空间则有机地布置于各“棋格”内。配合基地山体走势及用地形态，在规划布局上进行局部轴线偏移，形成两套独立的轴网系统，以提高基地实用率、增加建筑空间的丰富性，同时通过小中见大的空间走向、洁白的高墙等传统元素，从院落，到传统，到生态，到文化，力求营造出一个环境优雅、与自然和谐统一的棋院人文景观空间。

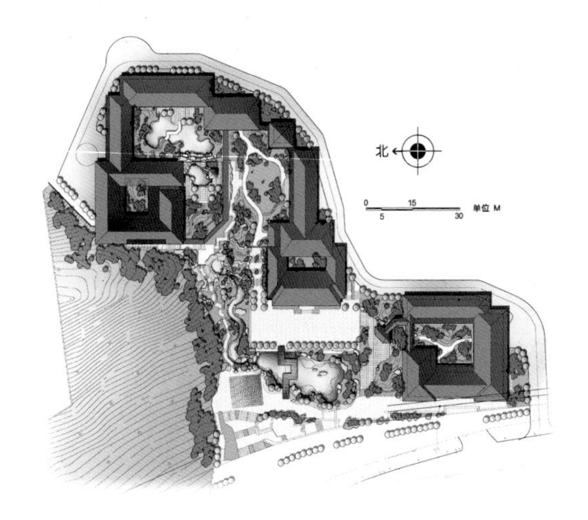

The Project is situated at Heng Zhi Gang Road in the Baiyun Mt Scenic Area, just to the east of the Lu Lake Park, Guangzhou. It is enclosed by the Baiyun Mt on the west and north, Guangzhou Air Force Military Land on the east and the land owned by Baiyun Mt Administration Office on the south.

The project design references the artistic atmosphere as depicted in an old Chinese verse going as "on a warm spring day, the door is closed in the tranquility of the daylight; among the thriving bamboo woods, I linger around leisurely and watch people playing chess". With the chessboard as the motif of the site planning, building volumes are separated from each other while interconnected via the flexible and changing walls. The building functions and courtyard spaces are organically placed in "squares" of the chessboard. To match the mountain extension and land use pattern of the site, the planning axis is partially shifted to derive two independent sets of axial network systems to enhance the land use efficiency and diversity of the building spaces. Meanwhile, such traditional elements as the spatial extension of "imaging spectacular view from small detail" and high white walls are employed to create an elegant and human-oriented space that co-exists and harmonizes with the nature, either in terms of the courtyard, tradition, ecology or culture.

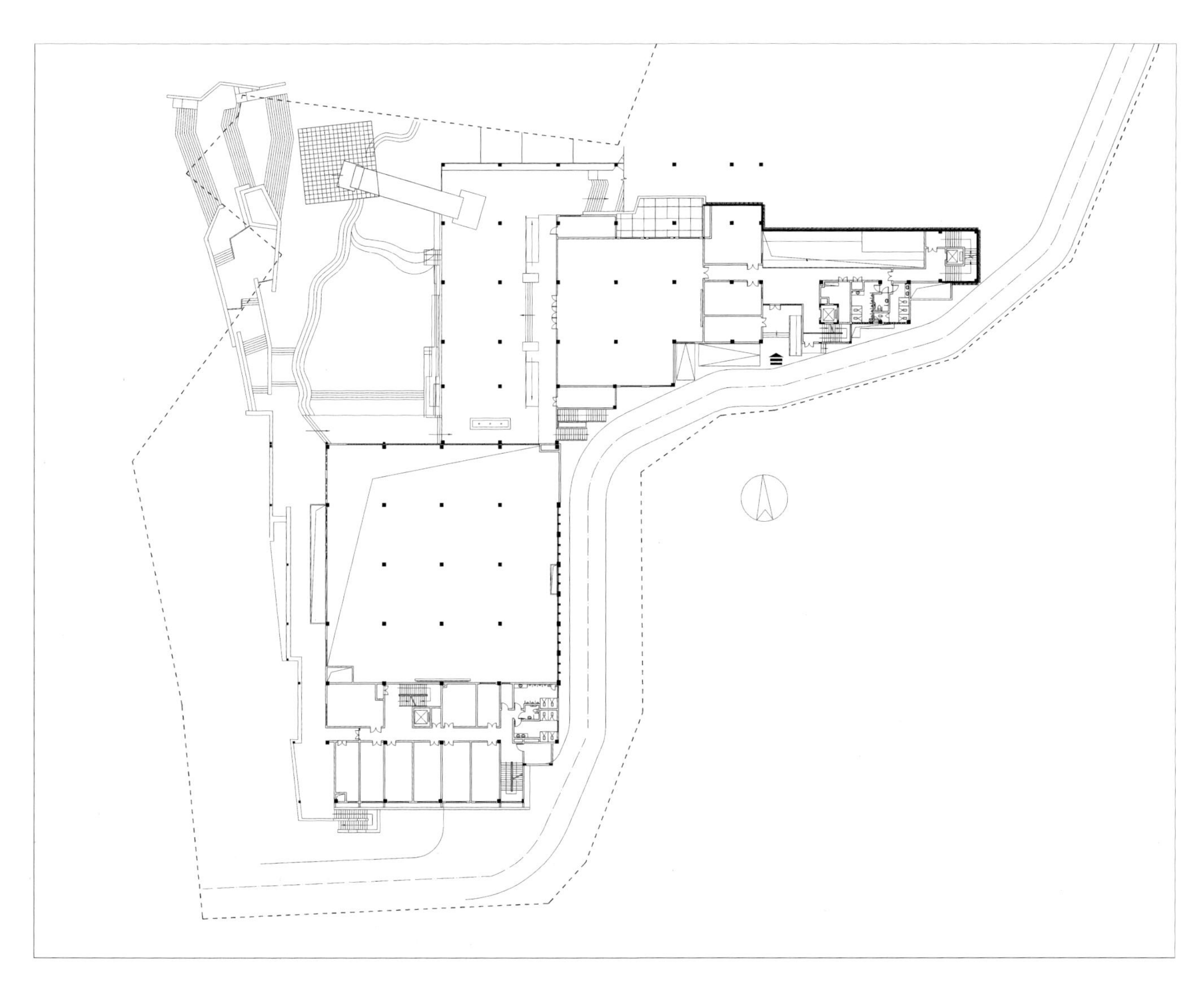

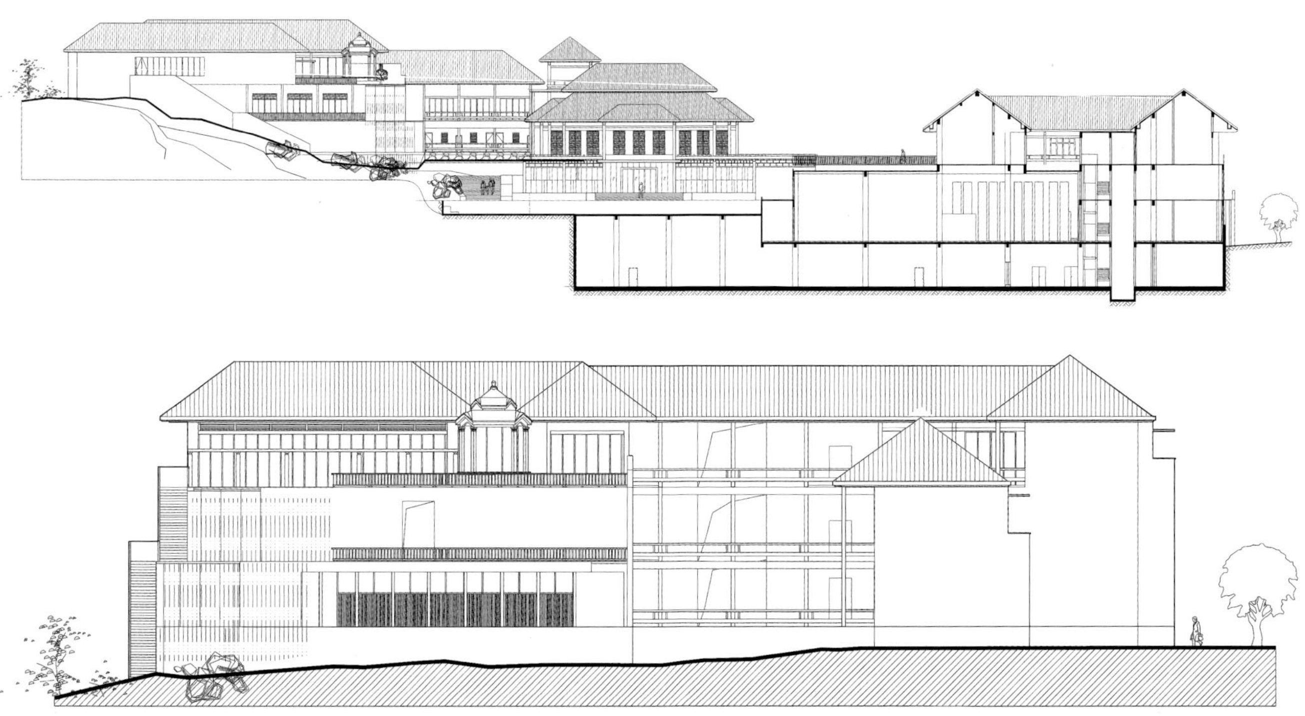

上至下：4.5m标高平面图、剖面图、立面图。
Top down: 4.5m Level Plan, Section, Elevation.

立面图 Elevation

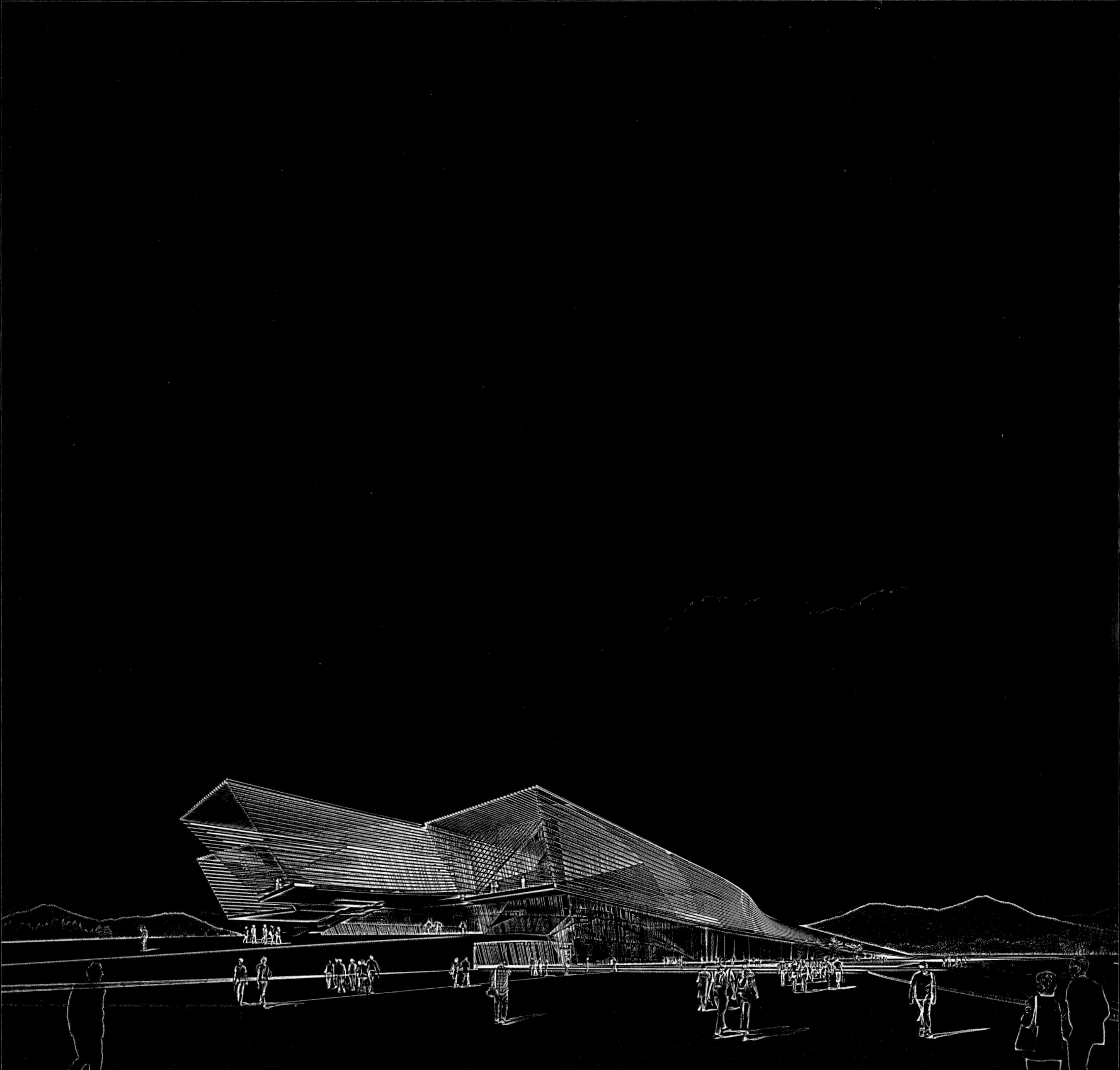

SCHEME DESIGN
方案创作类

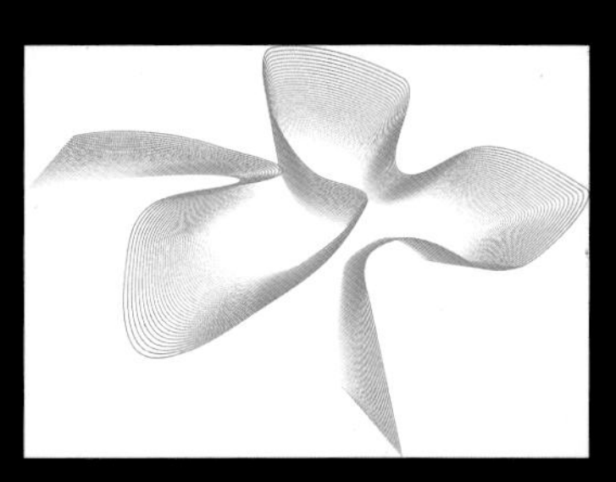

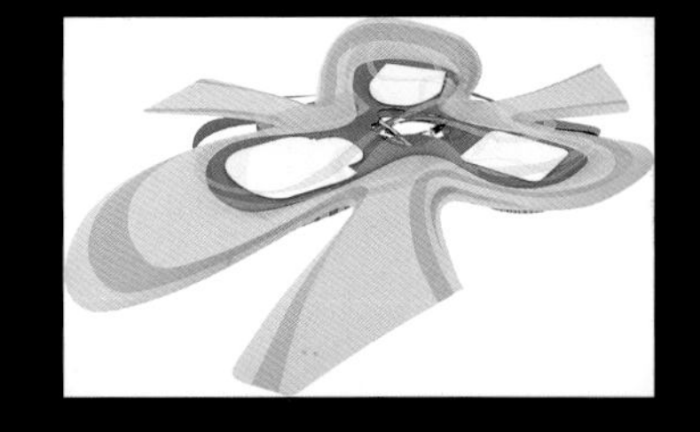

30 襄阳大剧院
Xiangyang Grand Theater

襄阳大剧院项目位于襄阳市东津新区西北角汉江与唐白河交汇之处，三面环水，地理位置和景观资源十分优越，是襄阳新城市中心区的焦点所在，是襄阳市的标志性文化建筑。襄阳大剧院地上建筑面积约为 5 万平方米，包括 1600 座大剧院、800 座音乐厅和多功能厅、报告厅、大中小会议室及文化配套服务设施。

老子曰："道生一，一生二，二生三，三生万物。"襄阳大剧院的建筑形态借用"三生万物"的生发之美，寓意襄阳之花、文化之花、生命之花。意即其在历史的

Xiangyang Grand Theater is located at the intersection of Hanjiang River and Tangbaihe River in the northwest corner of Dongjin New Area, Xiangyang City. It is surrounded by water on three sides. With advantageous geographical location and superior landscape resources, it marks a focal point of Xiangyang's new urban center and a cultural landmark in the city. With an above-grade floor area of about 50,000m^2, the Theatre houses a 1,600-seat grand theater, a 800-seat concert hall, as well as multi-purpose hall, lecture hall, small/medium-sized/large conference room, and supporting cultural facilities.

As Lao Tzu says: "The way bears sensation, sensation bears memory, sensation and memory bear abstraction, and abstraction bears all world". Inspired by the charm of growth as reflected in "abstraction bears all world",

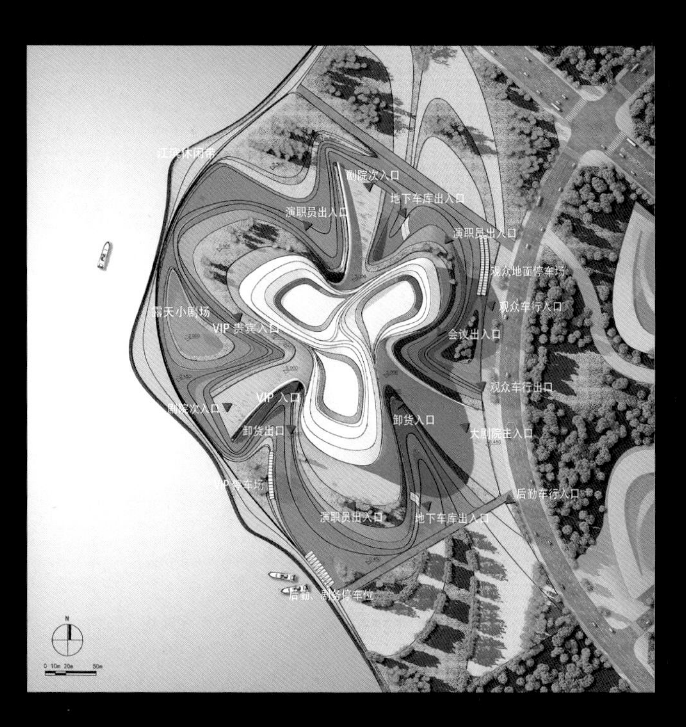

时空长河中绚丽地绽放，承前启后，生生不息。襄阳大剧院地处汉江与唐白河交汇之处，恰恰有如站在历史与未来的时空交点的我们，肩负着传承物质、精神文明的重任。

运用现代手法对歌舞剧的动态元素与汉水灵动之舞进行抽象提炼，飘逸优美的建筑形象与汉水灵动的自然形态瞬间组合，以动感的肌理形成互动的城市艺术文化

Xiangyang Grand Theater is designed in an architectural form that implies the flower of Xiangyang, of culture and of life blooming in the long process of history. The Theater standing at the intersection of Hanjiang River and Tangbaihe River is like us assuming the heavy responsibility of inheriting the material and spiritual civilization at the interchange of the past and the future.

The dynamic elements of musical dramas and the spiritual dance of Hanshui River are abstracted and refined via modern approaches; while dynamic fabric deriving from the instantaneous mix of graceful architectural image and smart and natural form of Hanshui River creates interactive urban spaces for artistic and cultural performances.

31 增城大剧院
Zengcheng Grand Theater

增城大剧院项目用地位于新城挂绿湖北岸，西临行政中心，北面与少年宫、体育馆等新建建筑并立，剧院在城市中起到连接北面城市片区与挂绿湖的作用。

大剧院的设计以一水一山为主题，注重建筑与环境的结合，设计保留了项目地块西面的山体，与现有的绿化广场形成一个整体，尽可能减少对现有绿化广场景观的影响。以河流环绕山流动的形态，将项目基地分为生态保留区和新建区两部分，建筑与景观相呼应。

人的动线规划参照水流的形态，曲韵交错，如同一股股水流汇入挂绿湖。主体建筑的形态亦跟随流线而走，如同何仙姑的飘带落至挂绿湖畔，跟随音乐飘舞，优美灵动，也犹如江水蜿蜒奔涌，汇入挂绿湖。

The site is located on the north bank of Gualyuhu Lake in the New Town area. It neighbors the Administrative Center on the west, and stands side by side with the Youth Palace and the Gymnasium on the north. The Theater also links the northern urban district to Gualyuhu.

The lake and mountain-themed design integrates the architecture with the environment. The mountain on the west of the site is retained to merge with the existing greening square with minimal impact on the landscape of the Square. The site plan is a metaphor of a charming image of rivers flowing around the mountains, and the site is therefore divided into two parts, i.e. the ecological conservation area and the new development area where buildings echo with the landscape.

The pedestrian circulation references the water flows rushing rhythmically into Gualyuhu Lake. The form of the main building also follow such flowing lines, just like a ribbon of Immortal Woman He landing the riverside and dancing airily to music, or a meandering river rushing down into Gualyuhu Lake.

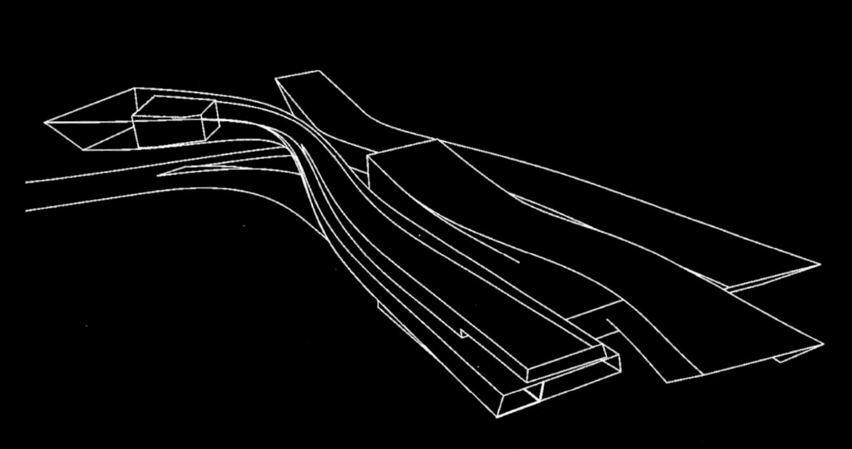

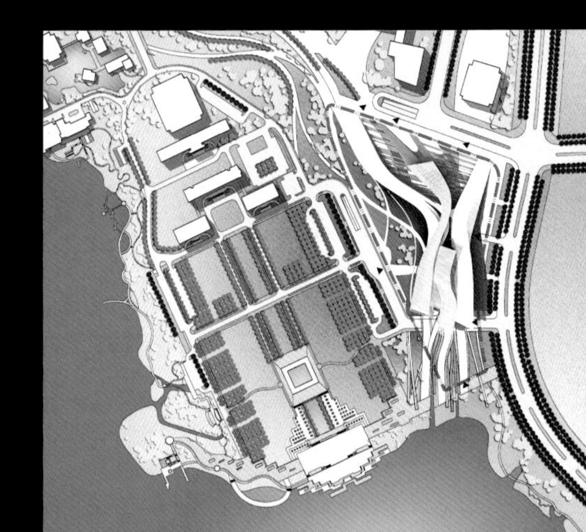

32 珠海歌剧院
（与扎哈・哈迪德建筑师事务所合作设计）
Zhuhai Opera House
(In Collaboration with Zaha Hadid Architects)

野狸岛地区概念规划通过文化艺术产业的共同打造，以艺术文化旅游圈、大剧院文化艺术中心演艺综合体概念，增强文化旅游的吸引力，进一步提升珠海城市中心区的整体艺术文化氛围，打造高雅文化平台、市民休闲载体、旅游观光亮点、城市标志景观。

项目用地约为 5.0 万平方米，其中大剧院建筑面积约为 3 万平方米，含 1500 座大剧院及 400 座音乐厅，拟建剧院配套文化休闲设施共约 3.5 万平方米，含美术文化展览馆、五星级酒店会议中心、文化艺术街。

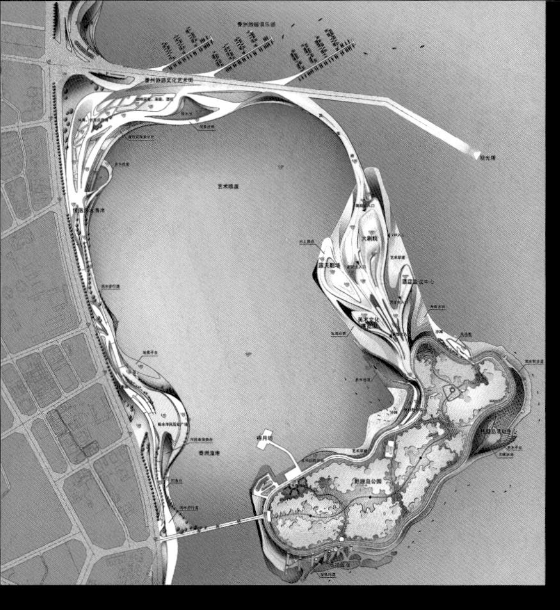
香洲游艇俱乐部
香洲旅游文化艺术街
大剧院
露天剧场
酒店会议中心
野狸岛公园

33 云南文化艺术中心
Yunnan Cultural and Art Center

云南文化艺术中心项目一期总建筑面积约为3.2万平方米，包括1500座大剧院、400座多功能厅、艺术文化展览及文化配套服务设施。

云南文化艺术中心提取滇人舞蹈形体动态元素，飘逸优美的建筑形象与大地舞动姿态的瞬间组合，以动感的肌理形成互动的城市艺术文化表演空间。建筑形体构成采用编织的手法，以锦带的延展、上升、扭动、穿插、自然编织，形成大剧院的独特、鲜明的建筑形体，宛如室内殿堂的歌舞旋转扰动着绣带翩翩起舞，源远流长。

该项目包括室外露天剧场、剧院主题文化广场、七彩梯田景观广场及七彩花带。建筑体量最终融入大地，和地景形成统一的总图肌理。将景观元素渗透到建筑形体和空间中，以动态的建筑空间完成功能与体量的分合，形成文化和景观的融合和共生。

Yunan Culture and Art Center (Phase I) in a GFA of about 32,000m^2 includes a 1,500-seat grand theater, a 400-seat multi-purpose hall, as well as some art and cultural exhibition and cultural supporting facilities.

By extracting the kinetic element of dancing Yunnan people, the project design realizes instantaneous integration between the grateful architectural image and the undulating landform, creating an interactive artistic and cultural performance spaces with dynamic building skin texture. Imitating the traditional weaving process, the ribbons are extended, uplifted, and twisted and intervened to create the unique and distinctive architectural form of the Grand Theatre. It seems that, along with the rhythms of the dance and music shows within the Grand Theatre, the ribbons start to move, rotate and wave elegantly and ceaselessly.

The project also includes open-air theater, theater-themed cultural square, colorful terrace landscape square, and multicolored flower strips. The building volume is finally merged into the earth, presenting a consistent fabric as the landscape in the master plan. Moreover, the landscaping elements are incorporated and penetrated into the architectural form and space, realizing the combination and subdivision of functions and volumes through the dynamic spaces and eventually realizing the integration and symbiosis of culture and landscape.

广 场 路
主入口
观众车行出入口
入口广场
观众地面停车场
露天小剧场
中心文化广场
剧院主入口
博物馆新馆
（一期）
博物馆入口
观众厅平台
演职员入口
后勤辅助入口
剧院二期文化广场
七彩梯田景观广场
文体艺术中心
博物馆二期
次入口

34 上海嘉定保利文化商业中心
Jiading Poly Cultural and Commercial Center, Shanghai

嘉定保利剧院及文化商业中心用地面积共5.6万平方米，总建筑面积约为14.2万平方米（含地下室），包括1500座的大剧院、400座的多功能厅、配套餐厅、演出技术用房等剧院配套用房以及完善的四星级酒店办公中心、电影院、会议中心、商场等文化商业中心。

总体构思提炼于长三角良渚文化、吴越文化、嘉定独特的水乡特色与民俗文化——江南社戏，并以民间传统艺术"嘉定竹刻"作为建筑的语言进行表达。剧院及文化商业中心将嘉定新城理念与固有的东方城镇古韵结合，在城市风貌上演绎了时尚现代与水乡风韵相互交融的乐章。时代与历史握手，时尚与古朴对话，在传承历史的基础上，充分体现了嘉定新城的无穷活力。

The Project is planned with a site area of about 56,000m^2 and a GFA of about 142,000m^2 (including the basement). It houses a 1,500-seat grand theater, a 400-seat multi-purpose hall, a supporting dining hall, and theater-supporting facilities such as technical rooms. There's also a cultural and commercial center with a self-contained 4-star hotel & office center, a movie theater, a conference center and a marketplace.

The design is inspired from Liangzhu Culture and Wuyue Culture in the Yangtze River Delta, the unique water village feature of Jiading, as well as the folk culture of Jiangnan Shexi (a village opera in regions south of the Yangtze River). Meanwhile, the traditional folk art, Jiading Bamboo Carving, is also employed as the architectural vocabulary. The Theater and the Cultural and Commercial Center integrate the design concept of Jiading New Town with the inherent rhythm of the ancient Eastern town, playing a movement where fashion and modernity are in complete harmony with the charm of the water village in the cityscape context. Here, the time meets with the history while the fashion conversates with the primitive simplicity. By inheriting the history, the infinite vibrancy of Jiading New Town is fully reflected.

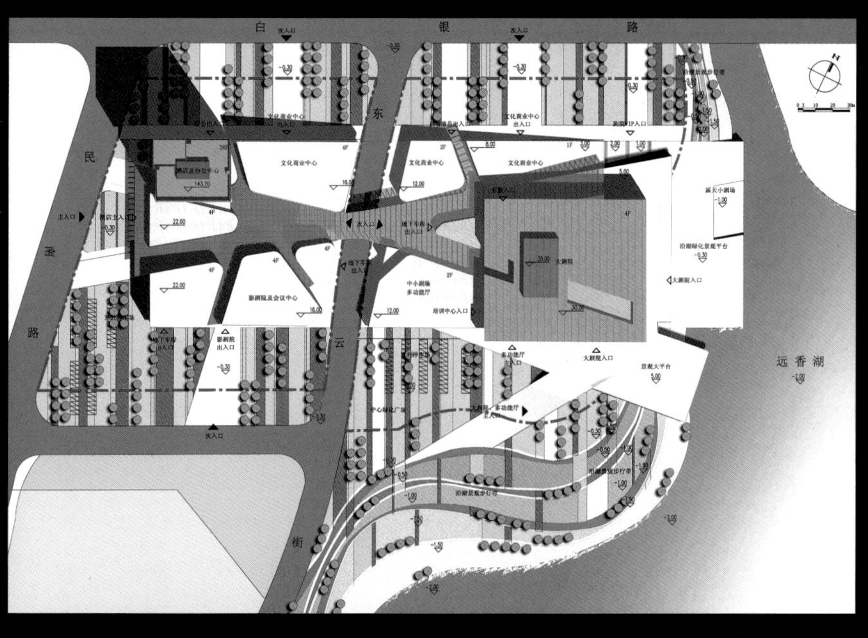
白银路
民南路
东云街
远香湖
次入口
主入口
文化商业中心
文化商业中心出入口
剧院VIP入口
酒店及办公中心
酒店主入口
影剧院及会议中心
影剧院出入口
地下车库出入口
中小剧场
多功能厅
培训中心入口
大剧院
大剧院入口
露天小剧场
沿湖绿化景观平台
景观大平台
多功能厅主入口
中心绿化广场

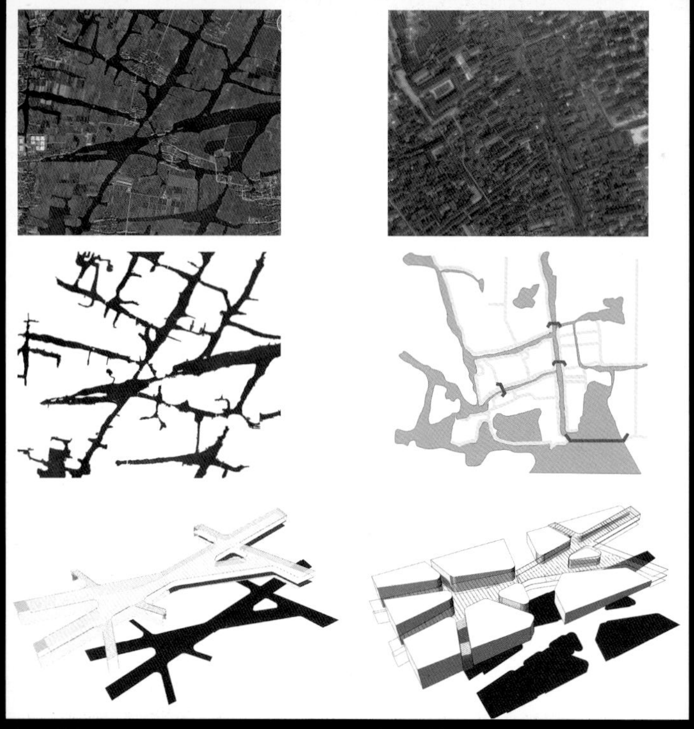

水系形态
Morphology of canal system

水乡形态
Morphology of water village

剧院 Theatre

多功能厅 Multi-purpose hall

酒店及会议中心 Hotel and Conference Center

办公楼 Office building

影剧院 Movie theatre

文化商业中心 Cultural and commercial center

立面的构成：
Facade composition

水网的形体，竹刻的手法
Presenting morphology of canal system through the bamboo-carving technique.

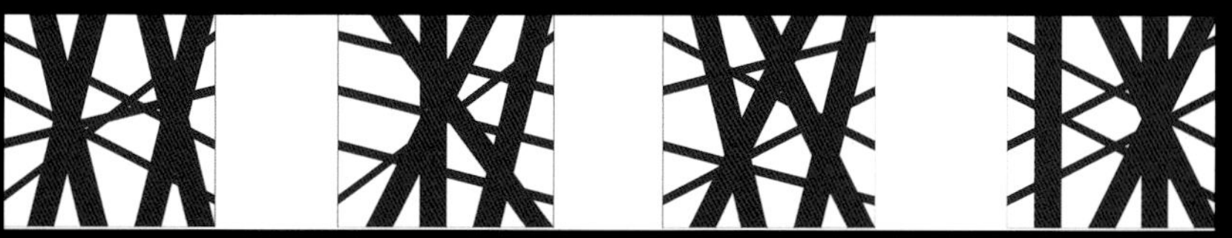

四个基本的预制混凝土单元
Four basic pre-fabricated concrete modules.

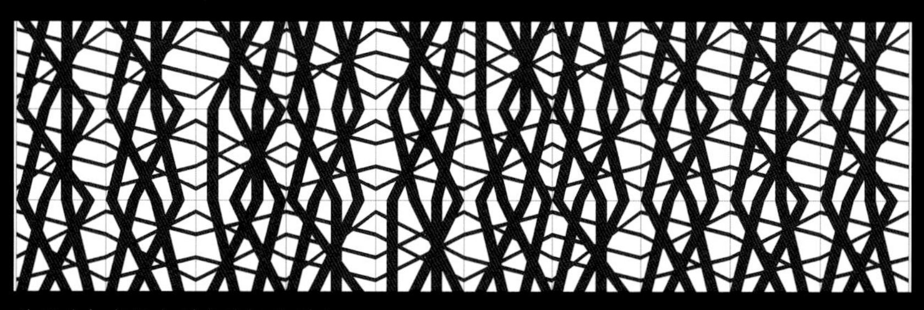

单元的任意组合形成随机的形态
Randomly combined modules show random patterns.

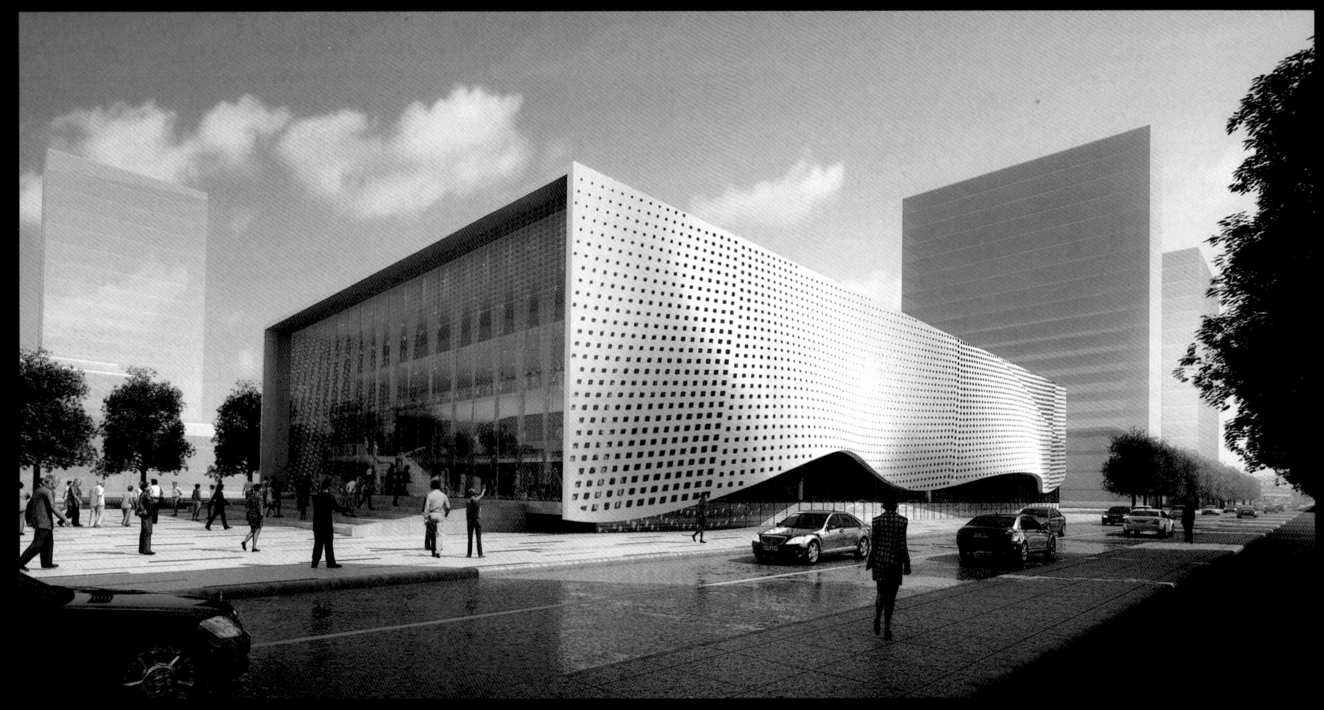

35 大连金州大剧院
Jinzhou Grand Theater, Dalian

城市舞台——大剧院是歌剧、芭蕾舞剧和大型文艺节目的文化表演场所，在都市之中，它又是市民共同拥有的享受或演绎生活艺术的舞台。大连金州大剧院面向城市的界面引入了舞台中帷幕的元素，寓意城市舞台上的帷幕。建筑入口如缓缓拉起的帷幕，暗示着这里是即将上演市民活动的"生活剧场"。大剧院的两个侧面又如轻轻翻起的幕布，剧院内部的光线从幕布下面渗出，让剧院充满了神秘的魅力。

The Theatre is to serve as an urban stage, for such cultural performances as opera, ballet and large theatrical shows to put on, and for citizens to enjoy or interpret the art of living within a urban context. The element of stage curtain is introduced to the interface of the Theater facing the urban area, implying the curtains for the *Stage of City*. The entrance to the building resembles a slowly raised curtain to suggest a forthcoming *Living Theatre* for citizen activities. The two flanks of the Theatre resemble the slowly turned-up stage curtains. The interior lighting penetrating from the bottom of the curtains adds mystery and charm to the Theatre.

石家庄市霞光大剧院（演艺中心）建设项目概念性方案设计
The Conceptual Scheme of Shijiazhuang Xiaguang Theatre (Performing Arts Center)

36 石家庄市霞光大剧院
Xiaguang Grand Theater, Shijiazhuang

传统曲艺如"河北梆子"等在民间流传的方式是"走街串巷"，或于集市撂地卖艺，或于庙会、庭院设棚。曲艺表演深入大众生活，临时搭设的"戏台子"往往氛围浓厚、热闹非凡，具有强烈的市民参与性。

设计通过传统"戏台子"这一民间特有的大众表演场所的再创造，形成一个个以"戏台子"为中心的演艺广场，作为建筑的室外延伸，最终形成气氛活跃、热闹非凡的城市空间，让剧院在没有演出的时候亦能充满生机。

The traditional Chinese folk art, such as Hebei Clapper Opera, is spread among the people by street tours, or street or shed (that is put up in a temple fair or a courtyard) performances. Chinese folk art shows are deeply rooted in the life of the general public. Temporarily set *stages* are normally festive and bustling and widely participated by residents.

Through recreation of the *traditional stage*, a unique place among the people for public shows, the design intends to establish an array of stage-centered performance squares as the exterior extensions of the building. These extensions will eventually create dynamic and lively urban

裕兴派出所
人行主出入口
6F 快捷酒店
平台
演艺广场
音乐厅入口
3F
音乐厅
大剧院
演艺广场
4F
6F
戏台子
演艺广场

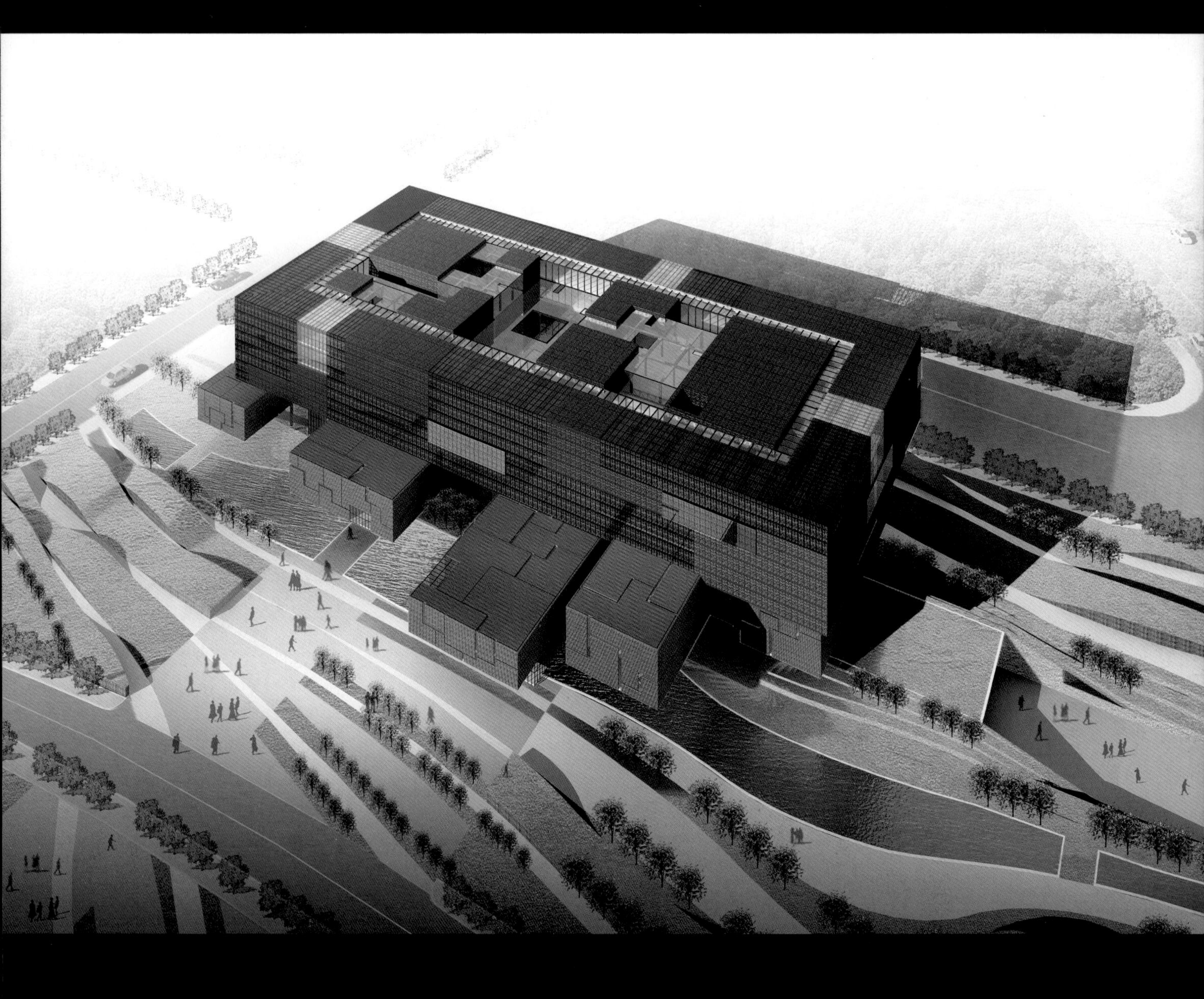

37 湖北省图书馆新馆
Hubei Provincial Library (New)

湖北省图书馆方案设计，就像一个永恒的藏书宝盒将源远流长的“楚文化”永久珍藏。建筑形体犹如一个飘浮在空中的藏书宝盒，主体下几个红色小宝盒高低错落，空间穿插有序，展示了大气、典雅的雕塑形象，虚实的组合变化交错形成多维度、多层次的缓冲空间，达到步移景异的效果。为了保持南北方向视线的通透，我们在新馆的 1 ～ 4 层采用局部架空的设计手法，以多个条形体量的穿插来削弱整体的体量感，弱化其对行人观景的影响。

奖项：

广东省注册建筑师协会优秀建筑创作奖（未建）。

Hubei Provincial Library is like a box treasuring the long-standing Chu Culture. The main building is in the shape of a floating box, under which a few red boxes in varied heights are scattered and interspersed. While showcasing the grand and elegant sculpture-like architectural image, the voids and solids are combined and interlaced to create multi-dimensional and multi-level buffer spaces and offer different sceneries at different steps. To guarantee transparent view in the north-south direction, F1 to F4 of the building are partially opened up. A number of strip-like volumes are interspersed to mitigate the overall volume and its adverse impact on pedestrian views.

Award:

Excellent Architecture Creation Award of Guangdong Chapter of Association of Chinese Registered Architects (ACRAGD) (Unbuilt).

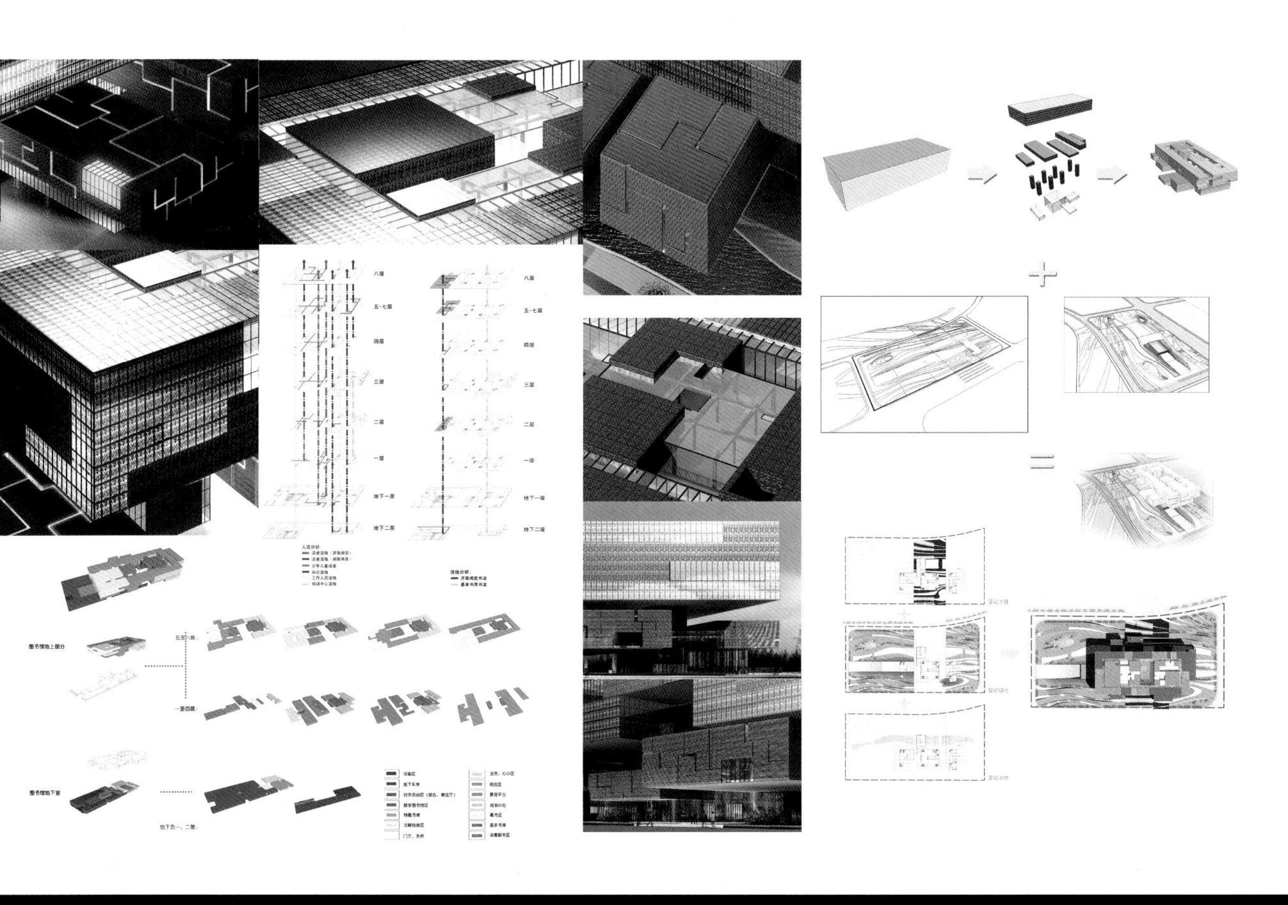

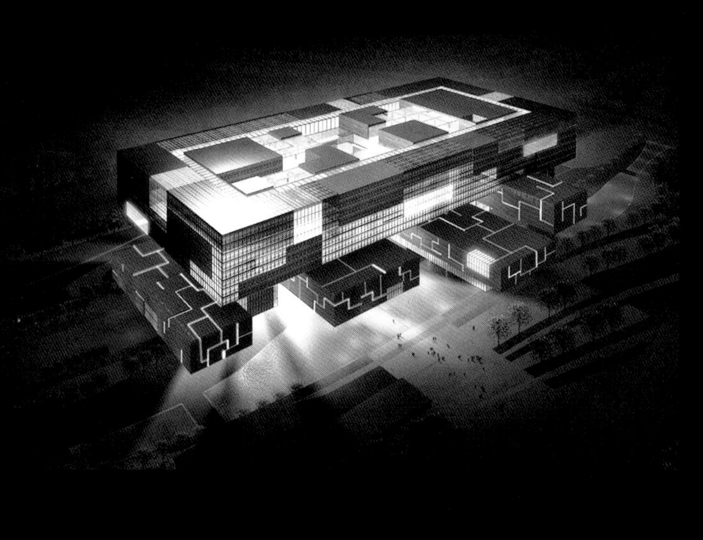

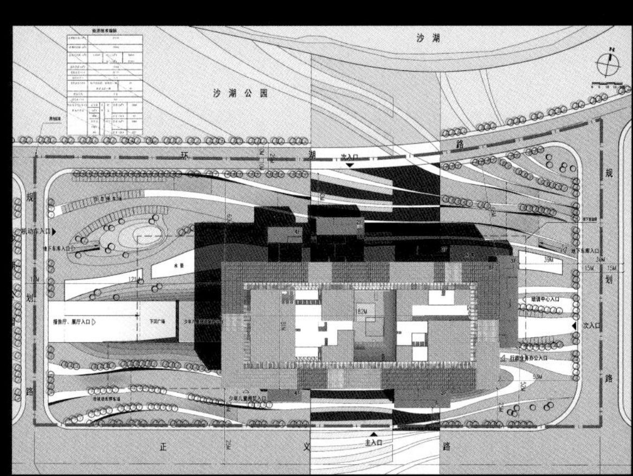

38 襄阳图书馆 Xiangyang Library

襄阳图书馆犹如一飘浮在空中的藏书宝盒，使用类似树木根系形象的清水混凝土剪力墙承托，表达文化之树的意象，从汉水流域吸取养分，孕育新的文化果实。刚柔并济是汉水文化中的一个核心思想，新图书馆的设计也以此作为一个出发点，在方正完整的建筑形体内部，以水流的形象，冲刷出“书的峡谷”，让读者在进入图书馆的瞬间就被书环绕着，产生知识殿堂的感觉以及阅读的冲动。

Xiangyang Library appears like a floating treasure box in the air. Fair-faced concrete shear walls resembling tree roots are employed to support the building, implying the *Tree of Culture* getting nutrients from Hanshui Basin to bear new *Fruits of Culture*. Combining hardness with softness is a core philosophy of Hanshui culture. Inspired by such philosophy, a "book canyon" is washed out by a "water flow" inside the regular and square architectural form. By this means, the readers will be surrounded by books the moment they enter the library, making the library a palace of knowledge and generating their interest in reading.

莫阳图书馆

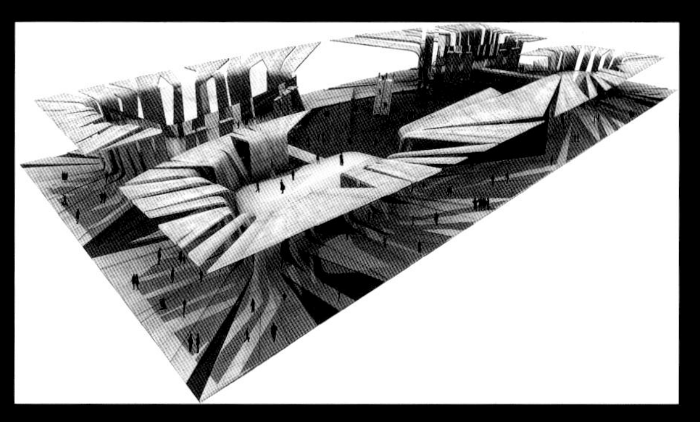

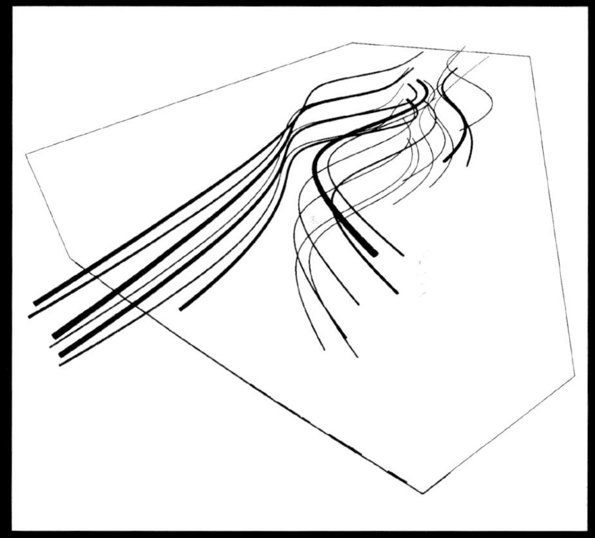

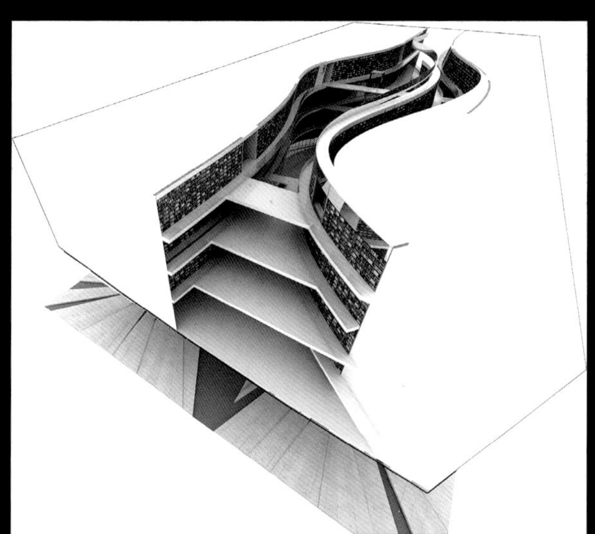

39 南越王宫博物馆
Nanyue Kingdom Palace Museum

本方案展示了本土生长而来的南越图腾发展到岭南特色的图腾文化的沉淀过程。南越时期发现的砖石走道上的菱形纹，其图案由比例多样的不同菱形以多种几何关系拼接。以菱形纹这个独具岭南地方特色的图腾作为母体，发展为博物馆的空间结构、展示空间和表皮的肌理。作为一层层系统，就如遗址一层层不同朝代演变沉淀一样，用图腾的演变延续在博物馆的外表与内涵中。

The design is an exhibition on how the indigenous Nanyue totem evolved to a totem culture with Lingnan characteristics. It is based on some diamond patterns found on the masonry walkway built in the Nanyue Kingdom period. The patterns on the walkway are made up of diamonds of various proportions put together by means of different geometric relations, and is considered as a unique totem with Lingnan characteristics. Such pattern is employed in the project design as a matrix for the growth of spatial structure, display space and skin fabric of the Museum. The layers of systems are like the layers of different dynasties evolving and sedimentating on a historic site, being continued in the appearance and connotation of the museum by means of totem evolution.

40 襄阳规划展示馆
Xiangyang Planning Exhibition Hall, Hubei

纵观襄阳的历史，它因文化而繁荣。温婉的汉江流经这一片土地，自三国时期，襄阳便汇聚了大量南北上下的文人，开创了一个思想学术与文化的高潮。其精髓是"天时"、"地利"、"人和"，得其三，则平天下。襄阳城市规划展示馆正是在这样的人文和地理环境下酝酿诞生的，作为一个城市展示馆，以展现襄阳的魅力为目标。

方案中建筑体块构成了一个类似窗口的形态，营造出"视窗"的形象。透过这样巨大的视窗空间，可看到襄阳城市规划的不断发展。视窗的西面有宽广悠长的襄阳母亲河——汉江，东面有襄阳城市规划发展的新区，透过"视窗"的形象，解读襄阳城市的过去、现在和未来。下沉式广场与建筑主体在形态上形成耦合关系，以穿插的路径和宽大的台阶，隐喻时空的流转以及城市的活力，并以地景建筑的手法，展示出生生不息的襄阳城市脉动。

Xiangyang flourished as a cultural center throughout history. Since the Three Kingdoms Period, this land irrigated by the gentle Hanjiang River has been the crossroads where scholars from the North and the South met and thus created an upsurge of academic thoughts and culture. The fundamental of these thoughts is that whoever granted with the right timing, superior geographical location and talent pool would be the king of the country. It is within such a historical and cultural context that Xiangyang Planning Exhibition Hall is proposed to showcase the unique charm of Xiangyang city.

The building appears to be a "window" with a window-like form. One may observe the evolution of the Xiangyang's urban planning through such gigantic window space. The "window" neighbors the long and broad mother river of Xiangyang, i.e. Hanjiang River on the west, and the planned new urban district of Xiangyang on the east. Via such a "window", the past, present and future of Xiangyang are all interpreted. The sunken plaza and the main building formally respond to each other . The interweaving paths and generous steps metaphorize the time-space evolution and the vitality of the city. Landscape architecture approaches are employed to exhibit the generative city beat of Xiangyang.

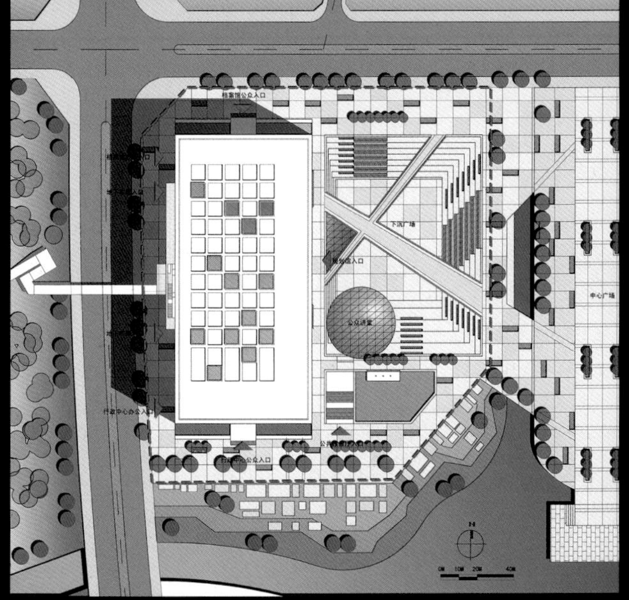
档案馆公众入口
下沉广场
规划馆入口
公众讲堂
中心广场
行政中心办公入口
行政中心公众入口
0M 10M 20M 40M

41 广州市国家档案馆新馆二期工程
Guangzhou National Archives (New), Phase II

国家档案馆分三期建设，本次设计为二期工程，属单体建筑。每一份档案就像一扇窗口，开在其中，吸引我们靠近，满足我们的求知欲。在窗口的另一侧，建筑的内部，是否也有另外一双眼睛透过窗口在注视着我们，注视着这个世界，记录着我们经历的点点滴滴。“历史的窗口”就是我们的设计的立意——档案馆不仅是记录资料，传承记忆的载体，也是我们回顾历史，审视现在，展望未来的窗口。

The National Archives is developed by three phases. The Project under the design is Phase II, a singular building. Each archive is like a window attracting people to approach and satisfy their thirst for knowledge. One may wonder that, on the other side of the window in the interior, whether there would be another pair of eyes watching us, or even the whole world and recording what we've gone through in the past. Based on this, we established our design concept, i.e. "The Window of History", meaning that the Archives serves as not only a carrier for data recording and memory inheritance, but also a window for us to look into the past, the present and the future.

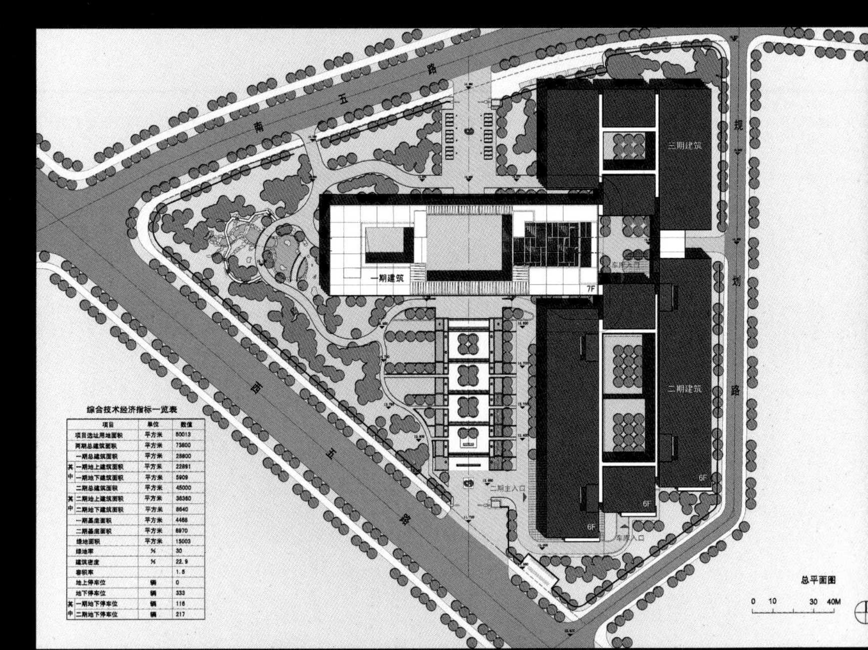

综合技术经济指标一览表

项目		单位	数值
项目选址用地面积		平方米	50013
两期总建筑面积		平方米	73800
一期总建筑面积		平方米	28800
其中	一期地上建筑面积	平方米	22891
	一期地下建筑面积	平方米	5909
二期总建筑面积		平方米	45000
其中	二期地上建筑面积	平方米	36360
	二期地下建筑面积	平方米	8640
一期基底面积		平方米	4468
二期基底面积		平方米	6970
绿地面积		平方米	15003
绿地率		%	30
建筑密度		%	22.9
容积率			1.5
地上停车位		辆	0
地下停车位		辆	333
其中	一期地下停车位	辆	116
	二期地下停车位	辆	217

总平面图

42 广州报业文化中心（与迈耶事务所合作设计）
Guangzhou Daily Group Cultural Center
(In collaboration with Richard Meier & Partners)

广州报业文化中心用地位于广州琶洲总部商贸区西北处 AH040102 地块，地处广州新城市中轴线南段东侧，北倚珠江，西望广州塔，东临珠江啤酒博览园，具有良好的区位优势。

广州南北新城区的城市肌理，确立了我们设计网格的方向，以此创建一个“广州报业文化中心”与城市空间秩序的有形链接。外立面材料运用尽可能精简，通过强调材料的轻盈和精密装配体现出“传媒”、“报业”的形象特质。

Guangzhou Daily Group Cultural Center is located on Plot AH40102 in the northwest of Pazhou Headquarters Trade Area, on the east of the south section of Guangzhou's new urban central axis. With the Pearl River on the north, the Canton Tower on the west and Zhujiang Beer Expo Park on the east, the site boasts advantageous geographic locations.

The urban fabric of the southern and northern new urban areas of Guangzhou defines the direction of our design grid. On such basis we intend to establish a visible link between the Project and the urban spatial order. The facade materials are minimized in design to highlight the lightness and precision assembling of the facade and portray a "media" and "newspapering" image for the Center.

43 兰州广播电视大厦
Lanzhou Radio and TV Tower, Lanzhou

通过对兰州地域文化的深层次发掘和研究，运用象征性的抽象形态以及具象的简化图形，对地域文化以及广播电视文化等进行了多方面的衍生和诠释，方案以稳重大气的建筑体形隐喻“兰州精神”，融合兰州人的性格。

建筑取义“黄河之波”，凸显黄河作为华夏文明发源地的意义；“像素之城”，凸显广播电视的特色；“魅力魔方”，给兰州城市带来意想不到的惊喜；“兰州视窗”，寓意广播电视大厦时刻关注着兰州城市发展和兰州市民的生活。“市民论坛”，传承兰州的包容性，尽可能为公众创造一个思想交流，让文化碰撞的场所，不断开拓创新，续写兰州新的伟大篇章。

Based on in-depth excavation and study on the local culture of Lanzhou, the local and radio & TV culture are derived and interpreted from various aspects in the design by means of symbolic and abstract form and figurative and simplified figures. A calm and generous building form is proposed to suggest the "Spirit of Lanzhou" and fit the disposition of Lanzhou people.

The architectural design elements comprise the "Wave of the Yellow River" that highlights the role of the River as the cradle for the Chinese civilization; the "City of Pixels" that addresses the radio and TV characteristics of the Tower; the "Rubik's Cube" that brings surprises to Lanzhou; and the "Window of Lanzhou" that implies the continuous attention paid to the city's development and local people's life. The "Citizens Forum" also seeks to inherit the inclusive spirit of Lanzhou. The design aims to create for the general public a place for brainstorms and cultural exchanges, so that they could continue a new great chapter for Lanzhou through constant exploration and innovation.

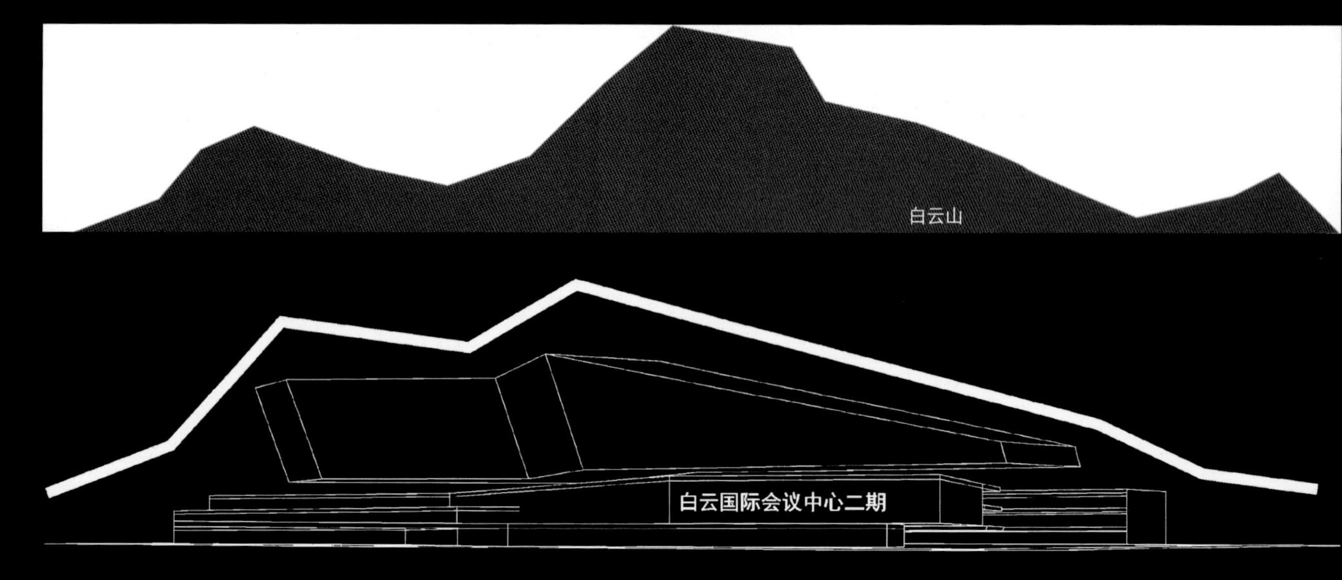

44 白云国际会议中心二期
Baiyun International Convention Center (Phase II)

白云国际会议中心二期场地位于广州市白云区白云大道以西、云城东路以东，南临广州白云国际会议中心，东面为白云山风景区。

白云山山势起伏绵延，自然地向四周延伸。白云山不应以白云大道为界，自然生态也不应只有山上有，更应延续到城市中去。我们将本案建筑群视为绵延的白云山的一部分，以山形为意，承接白云之势，与周边环境，特别是一期建筑群相融合，和而不同，共同创造和谐统一的城市新环境。

Baiyun International Convention Center (Phase II) is located in Baiyun District of Guangzhou City, west of Baiyun Avenue and east of Yuncheng East Road. It neighbors Baiyun International Convention Center on the south and Baiyun Mountain Scenic Spot on the east.

The rolling Baiyun Mountain stretching naturally toward the surroundings should not be bounded by Baiyun Avuen. The natural ecology should not just exist on the mountain but rather extend into the city. In our design, the building cluster of the Project is viewed as an integral part of Baiyun Mountain. The mountain shape is employed in the Project as extension of the magnificent Baiyun Mountain. By doing so, the Project is perfectly integrated into the surroundings, in particular the Phase I cluster, in a harmonized yet differentiated manner, and eventually create a harmonious and consistent new urban environment.

办公出入口
公寓出入口
办公
公寓
商业街出入口
商业内街
酒店
酒店主入口
宴会厅入口
酒店次入口
酒店主入口

45 大连国际会议中心（与扎哈·哈迪德建筑师事务所合作设计）
Dalian International Convention Center
(In Collaboration with Zaha Hadid Architects)

大连国际会议中心位于东港新区的西端，与城市主轴线毗邻，是整个“亲水海岸”及“地标性景观”的起点。建筑造型设计延续景观设计的语言以求两者的统一。其圆润且充满张力的形体试图表现如水流般的景观。向上延伸回旋的动态瞬间连续的带有拉伸和渐变效果的玻璃幕墙随势展开、延伸，在为建筑外形带来令人惊喜的艺术效果的同时，也将为建筑室内洒下神奇的光影。

The Project is located at the west end of Donggang New Area, adjacent to the city's central spine. It marks the starting point of the whole "water-accessible seashore" and "landscape landmark". The architectural form seeks to be consistent with the landscape by continuing the latter's vocabulary. Such smooth yet full-of-tension metaphorizes the free flowing landscape. Dynamically connected glass facade with stretching and gradient effects is then developed and extended to the building top in a revolving manner, bringing amazing artistic effect to the architectural form and magical light-shadow effect to the interior of the building.

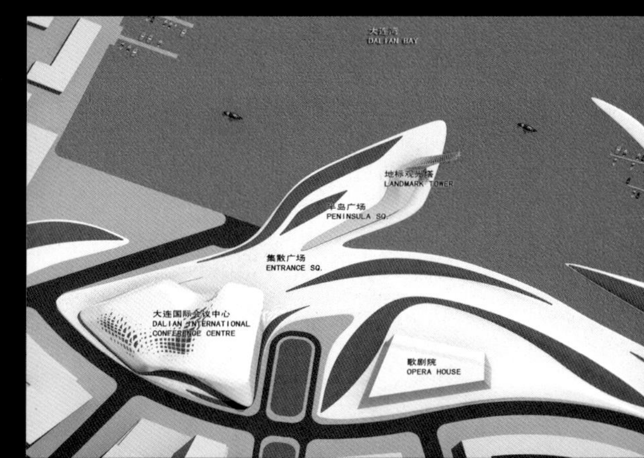
大连湾
DALIAN BAY
地标观光塔
LANDMARK TOWER
半岛广场
PENINSULA SQ.
集散广场
ENTRANCE SQ.
大连国际会议中心
DALIAN INTERNATIONAL CONFERENCE CENTRE
歌剧院
OPERA HOUSE

46 武汉市民中心
Wuhan Civic Center

武汉市市民中心作为一个开放性的城市公共空间，以其人性化的空间迎接到访的市民。建筑设计构思取自“白云黄鹤”，以抽象化的白云形态飘动于大地之上，未至黄鹤，先现白云，建筑以特有的诗意和现代的表现手法，体现了武汉千年来的文化积淀，映千古，向未来，成了武汉一道靓丽的城市风景！

奖项：

广东省注册建筑师协会 2009 年度创作奖。

As an open and humanistic public urban space, Wuhan Civic Center presents a welcoming gesture to all the visitors. The architectural design is conceived from the "white clouds and yellow crane", i.e. with abstract clouds floating above the ground before the yellow crane is yet to come. Distinctive architectural poetry and modern representation approach work together to showcase the cultural deposits of Wuhan during the past thousand years. Reflecting the past and looking into the future, the building will surely become another attractive view in Wuhan city!

Award :

Excellent Architecture Creation Award of Guangdong Chapter of Association of Chinese Registered Architects (ACRAGD), 2009.

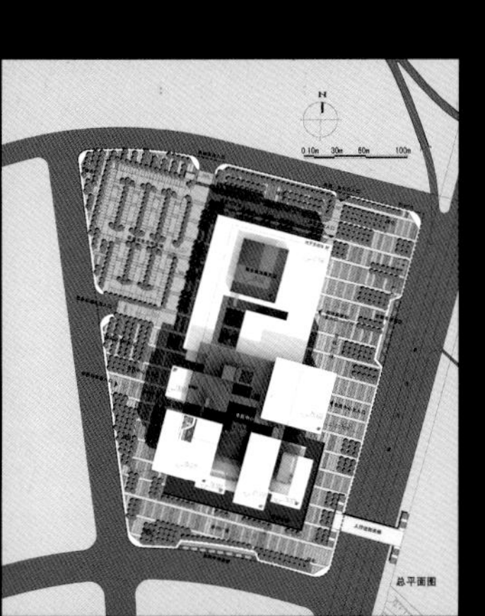
N
0 10m 30m 60m 100m
总平面图

47 襄阳市市民中心
Xiangyang Civic Center

襄阳是一座历史文化名城，是重要的交通物流枢纽和文化交流的通道，有“南船北马，七省通衢”之称。凭山之峻，据江之险，素来都是重要的门户城市。我们提出了“城市之门”的设计理念，寓意着通过市民中心将政府与市民联系在一起，把市民中心变成了展现城市的门户，沟通交流的平台，展望未来的窗口，在熙熙攘攘的广场人群中述说着政府与人民和谐共进的关系。

The historic city Xiangyang is an important hub for transportation and logistics, and a key route for cultural exchanges. It has long been known as "a meeting place for ships from the South and carriages from the North" and "a traffic hub leading conveniently to various provinces". Boasting steep mountains and torrential rivers, it has always been a vital gateway city. The design concept of "City Gate" implies the intent to link the government to the citizens via the Civic Center and turn the Center into a gateway for the display of cityscape, a platform for communication, and a window to look into the future. Between sunrises and sunsets, among the crowds on the bustling square, the Center vividly showcases the harmonious and mutual development between the government and the masses.

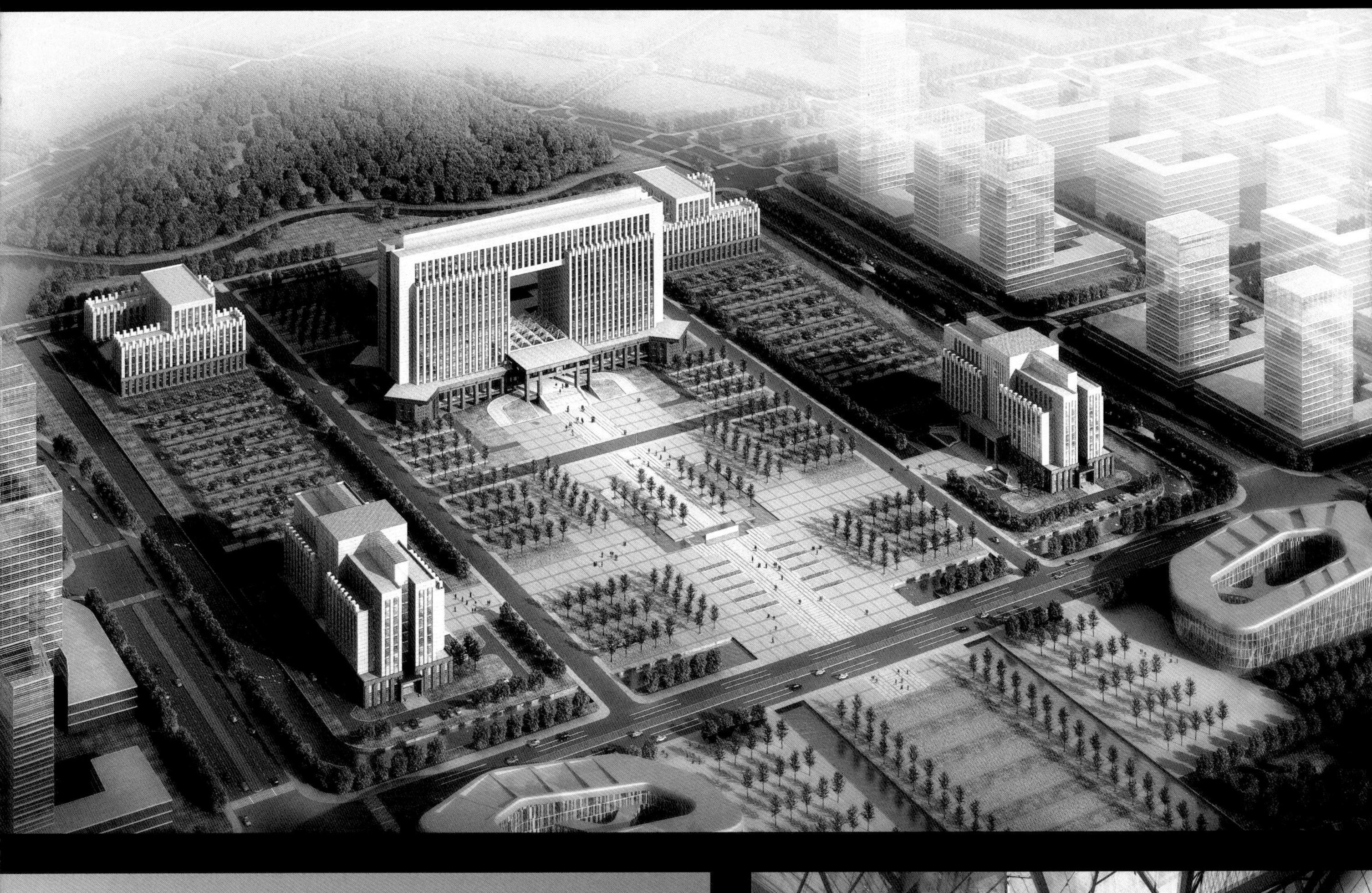

48 绵阳市职业教育园启动区第三标段
Kick-off Zone of Mianyang Vocational Education Park, Bid Section 3

本方案结合现代校园信息密集化的特点，把整个园区看成是一个灵活、动态的复杂性系统，运用复杂性科学的相关理论，通过定义一个建筑的原型，裂变出几种基本形态后再经过不同的组合方式排列组合出复杂的网络系统，建立起方案“自动生成”的过程。此方案的规划构思源于“细胞城市”概念，把每座建筑看成细胞，带有记忆细节的“基因”体现在每栋建筑（细胞）中。把这种概念应用在校园规划中，恰恰符合了学校信息密集、传输快捷和功能整合的需要。

In the design, the Park is considered as a flexible and dynamic complex system on the whole in consideration of the information-intensive characteristic of modern campuses. By applying relevant theories of complexity science, the design firstly defines a building prototype, then fissions a number of basic forms from such prototype, and then builds a complex network system with these basic forms via different arrangement and compound modes. The process of "automatic generation" is then established for the design. The planning is conceived from the concept of "cellular city", i.e. each building is considered as a cell accommodating genes that memorize all details. Applying such concept to campus planning has precisely met the demands of the campus for information intensification, convenient transmission, and functional integration.

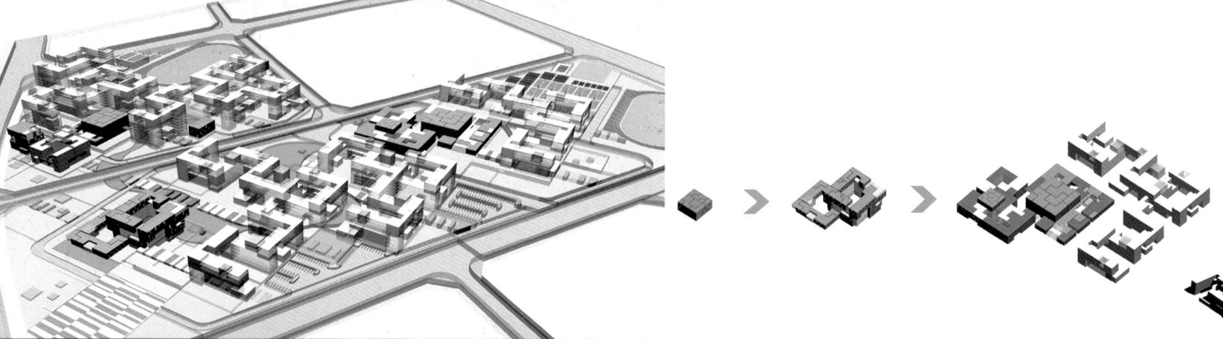

49 广东省天然气管网二期工程调控及应急指挥中心
Natural Gas Pipeline Network Project of Guangdong Province, Phase II Control and Emergency Command Center

本项目主体建筑有 80 米限高要求，在建筑形体设计上很难得到一个好的高宽比。因此，我们在外观设计上另辟蹊径，着力打造一个新颖、有趣、有寓意的标致性造型。

设计构思为三盏生动可爱的彩灯：寓意“张灯结彩”。三座塔楼犹如三盏彩灯，色彩艳丽，配以夜景灯效后更加绚丽多彩，活泼可爱，使人过目难忘。“彩灯”寓意广东天然气管网有限公司事业朝气蓬勃，生意兴隆，同时“灯火”的概念也与“气电”寓意暗合。

The max. height requirement of 80m for the main building made it difficult to work out a volume with decent height-width ratio. The design managed to find another way out from the aspect of appearance design, i.e. exerting all efforts to create a novel, interesting and moral landmark form.

The design proposes three lovely "colored lanterns", implying the meaning of being decorated with lanterns and festoons. The three towers are like three gorgeous colored lanterns and, coupled with the light effect at night, they look even cuter and more impressive. The "colored lanterns" is also an auspicious symbol for the business of Guangdong Natural Gas Pipeline Network Co., Ltd.; meanwhile, the "lights" concept also coincides with the implication of "gas and electricity".

50 广东省疾病预防控制中心
Centre for Disease Control and Prevention of Guangdong Province

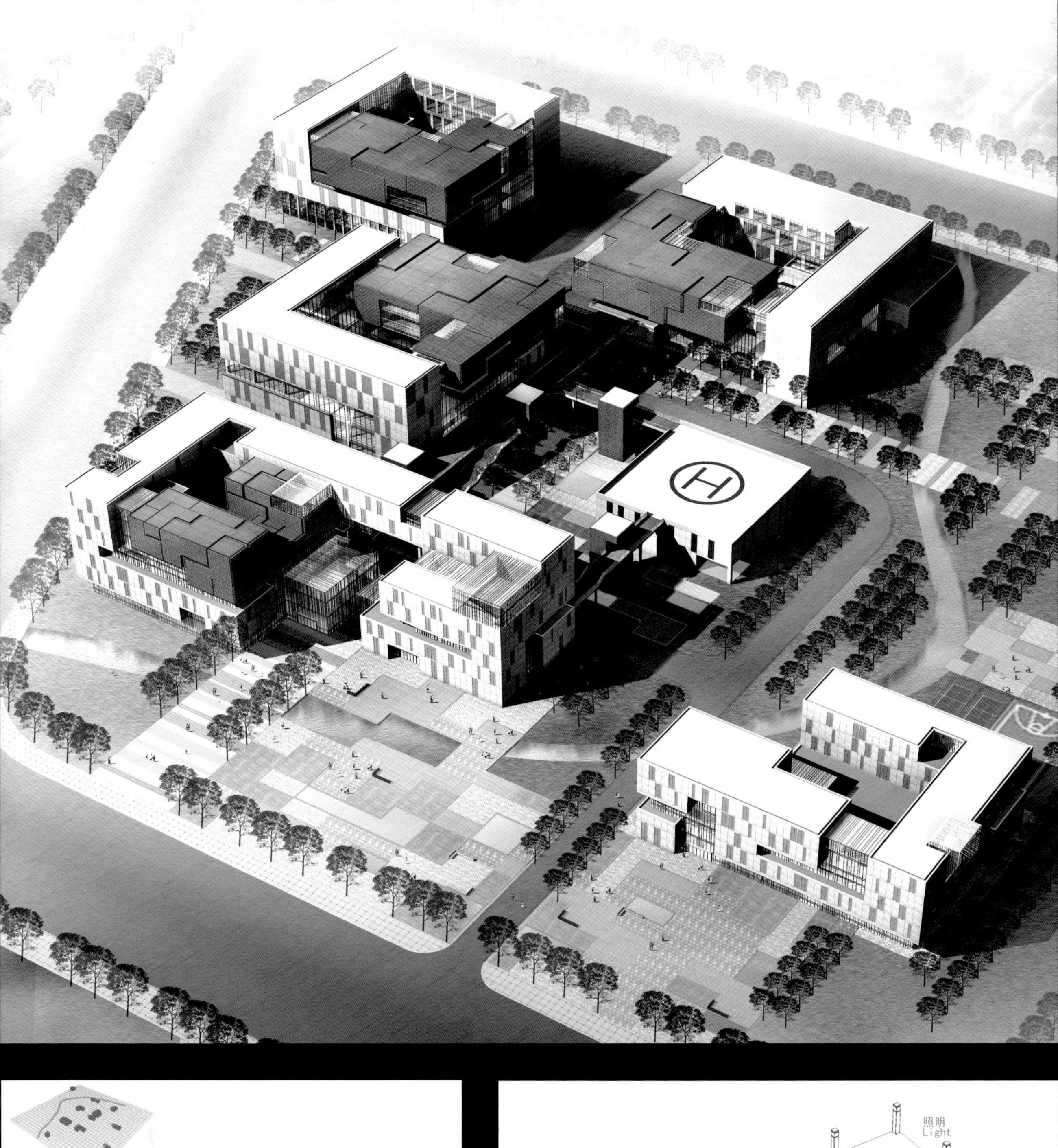

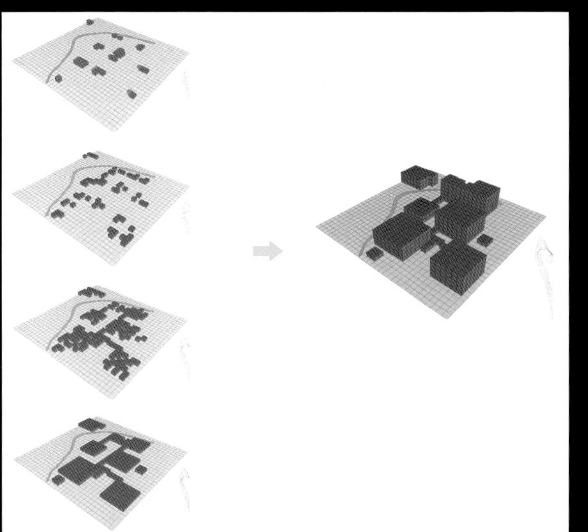

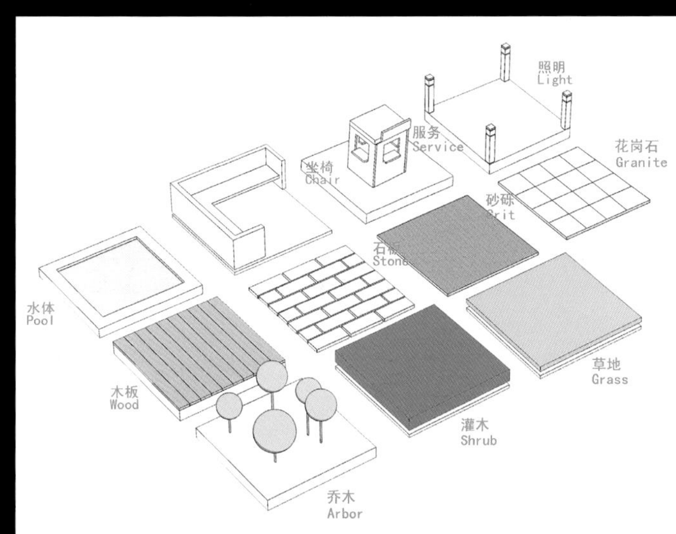
照明
Light
服务
Service
花岗石
Granite
坐椅
Chair
砂砾
Grit
水体
Pool
木板
Wood
草地
Grass
灌木
Shrub
乔木
Arbor

51 宁安线火车站房——弋江站、繁昌西站、安庆站

Station Buildings of Ning'an Intercity Railway — Yijiang Station, Fanchang West Station, and Anqing Station

南京至安庆铁路是长三角城际客运铁路网的延伸，地处华东皖南地区，居长江以南，铁路走向基本平行于长江，东起南京南站，终于安庆市，线路全长 257.48 公里。本设计方案的三个站为弋江站、繁昌西站及安庆站。

基于弋江四面临江，被“水”包含，因此立面抽象提取波浪纹理，突出弋江的地理特征。繁昌西站为了突出交通建筑特色，在建筑外立面采用铁路轨道横竖线条交错肌理，使立面富有交通的流动感。安庆站从结合当地徽派建筑特色入手，以简洁现代的手法，强调体量感，体现安庆站的现代建筑感，又不失徽派建筑韵味。

The Nanjing-Anqing Railway serves as an extension of the intercity passenger railway network in the Yangtze River Delta. It is located at the south of Anhui in South China, on the south of and basically in parallel with the Yangtze River. The railway starts from Nanjing South Railway Station and ends at Anqing City, totaling a length of 257.48km. The design covers three stations, i.e. Yijiang Station, Fanchang West Station, and Anqing Station.

Since Yijiang is surrounded by water at its four sides, the facade of Yijiang Station employs the abstract wave pattern to highlight such geographic conditions. In a bid to highlight its feature as a transportation building, Fanchang West Station uses interwoven horizontal and vertical lines resembling the railway tracks on the facade to present a sense of flowing transportation. For Anqing Station, based on the local Anhui-style architectural features, concise and modern approaches are adopted to highlight the sense of building massing and reflect the both contemporary and local (Anhui) style of the Station building.

弋江
弋江

繁昌西

繁昌西

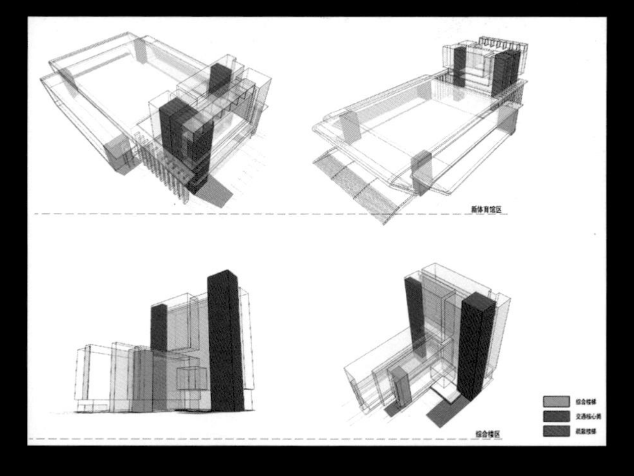

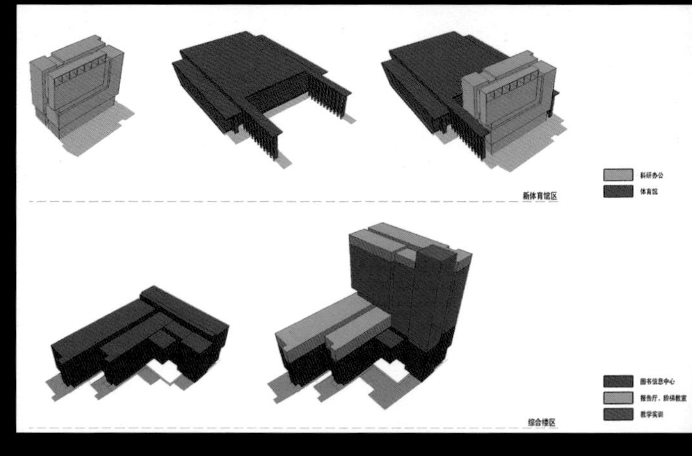

52 广州体育职业技术学院新体育馆及综合楼

New Gymnasium and Complex Building, Guangdong Vocational Institute of Sport

本方案首先对建筑体量进行分解，通过体量的错位创造多变的建筑形象，减少建筑对校园入口、道路的压迫感，展现现代校园轻松的特质。建筑通过统一的细部处理形成面向主入口的统一的建筑形象，同时，体块穿插和局部动感线条的灵活运用展现了动静皆宜的视觉品质，将建筑整体融入充满体育精神的校园风格中。

The design firstly deconstructs and shifts the building volumes to create a changeful architectural image, mitigate the pressure of the building on the campus entrance and road, and showcase the unique relaxing and lively environment on modern campus. Via unified treatment of details, a consistent architectural images is portrayed to face the main entrance; meanwhile, alternation of building volumes and dynamic lines at partial positions are flexibly employed to reflect the both active and passive visual quality, and incorporate the buildings on the whole into the campus full of spirit of sports.

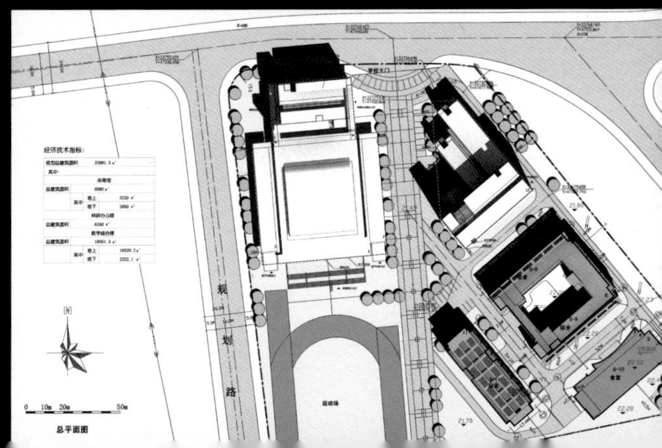

经济技术指标：
规划总建筑面积
其中：
体育馆
总建筑面积 9080 ㎡
其中 地上 5230 ㎡
地下 3850 ㎡
科研办公楼
总建筑面积 6350 ㎡
教学综合楼
总建筑面积 18551.3 ㎡
其中 地上 16329.2 ㎡
地下 2222.1 ㎡
N
规
划
路
运动场
学校大门
0 10m 20m 50m
总平面图

53 肇庆市体育中心
Zhaoqing Sports Center, Zhaoqing

肇庆市体育中心项目建设用地位于肇庆高新区中心城区，大旺大道东侧，北临曙光街，南面为政德街，东临经十一路。其中北面是市民滨海公园建设用地，东面为泛舟商业步行街，西南为商业办公建设用地。用地周边环境优美，西北面为景观山坡，东北面为幽静湖泊，自然景观丰富。

本设计以龙腾为设计依据，通过形态、结构、材质划分出正负空间构架，以工业设计手法隐喻龙腾虎跃、积极向上、永不言弃的体育精神。以曲线为母体，抽象刻画“龙腾之势”，简洁而富有象征性，营造出富有动感、视觉冲击强烈的标志性体育建筑，展示了肇庆市蓬勃发展、不断进取的城市活力。

Zhaoqing Sports Center is located at the central area of Zhaoqing New and Hi-tech Zone, neighboring Dawang Avenue on the west, Shuguang Street on the north, Zhengde Street on the south, and Jing Shi Yi Road on the east. The site borders on the construction land of Binhai Park on the north, Fanzhou Commercial Pedestrian Street on the east, and a commercial and office land on the southwest. With landscape slopes on the northwest and a tranquil lake on the northeast, the site boasts abundant natural landscape resources.

The basic concept is the "rising dragon". In the design, the frame for the positive and negative spaces are established as per form, structure and material, while industrial design approach is employed to imply the vigorous, positive, and never-give-up sportsmanship. Curve lines are adopted as the matrix to depict the magnificence of the "rising dragon" in an abstract manner. Such concise and highly representative lines lead to a dynamic sports landmark with strong visual impact, showcasing the vigorous development of the Sports Center and the enterprising urban vitality of Zhaoqing.

建筑退缩线
用地边线
次入口
主入口
次入口
次入口
大
旺
大
道
曙
光
街
政
德
街
N

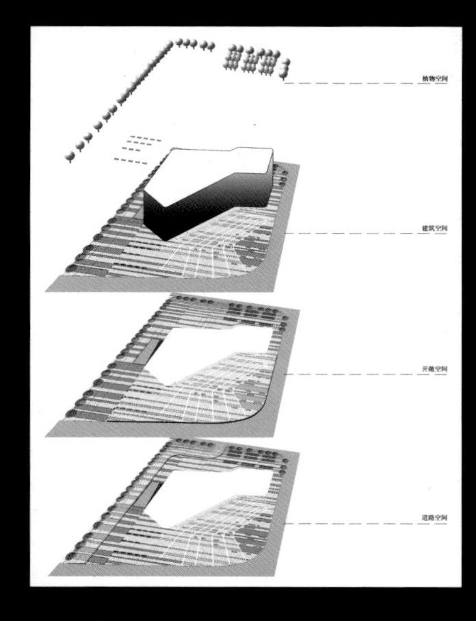

54 星海排练厅
Xinghai Rehearsal Hall, Guangzhou

星海集团新址建设用地面积为7325平方米，南面珠江，东邻星海音乐厅，环境优美，交通便利。本建设项目是为了解决星海演艺集团暂无排练场、办公用房的问题，为了促进广州交响乐的进步和发展而设的。

本设计通过看似不规则的表皮和景观来表现音乐创作的连贯性、流动性和不确定性，而建筑复杂、多变的室内空间更隐喻了乐章的序幕、开端、高潮和结局，同样的节奏通过不同的载体游走在空间里，飘扬在空气中。

With a site area of 7,325m^2, the new site of Xinghai Group faces the Pearl River on the south and Xinhai Concert Hall on the east, enjoying attractive environment and convenient traffic conditions. The Project aims to address Xinhai's shortage of rehearsal hall and offices and promote the advancement and development of Guangzhou's symphonic undertaking.

Seemingly irregular building skin and landscape are employed to represent the coherence, fluency and uncertainties of music creation, whereas complex and changeful interior spaces are provided to imply the prelude, beginning, climax and ending of a movement. Via different carriers, the same rhythms wander freely in these spaces and dance lightly in the air.

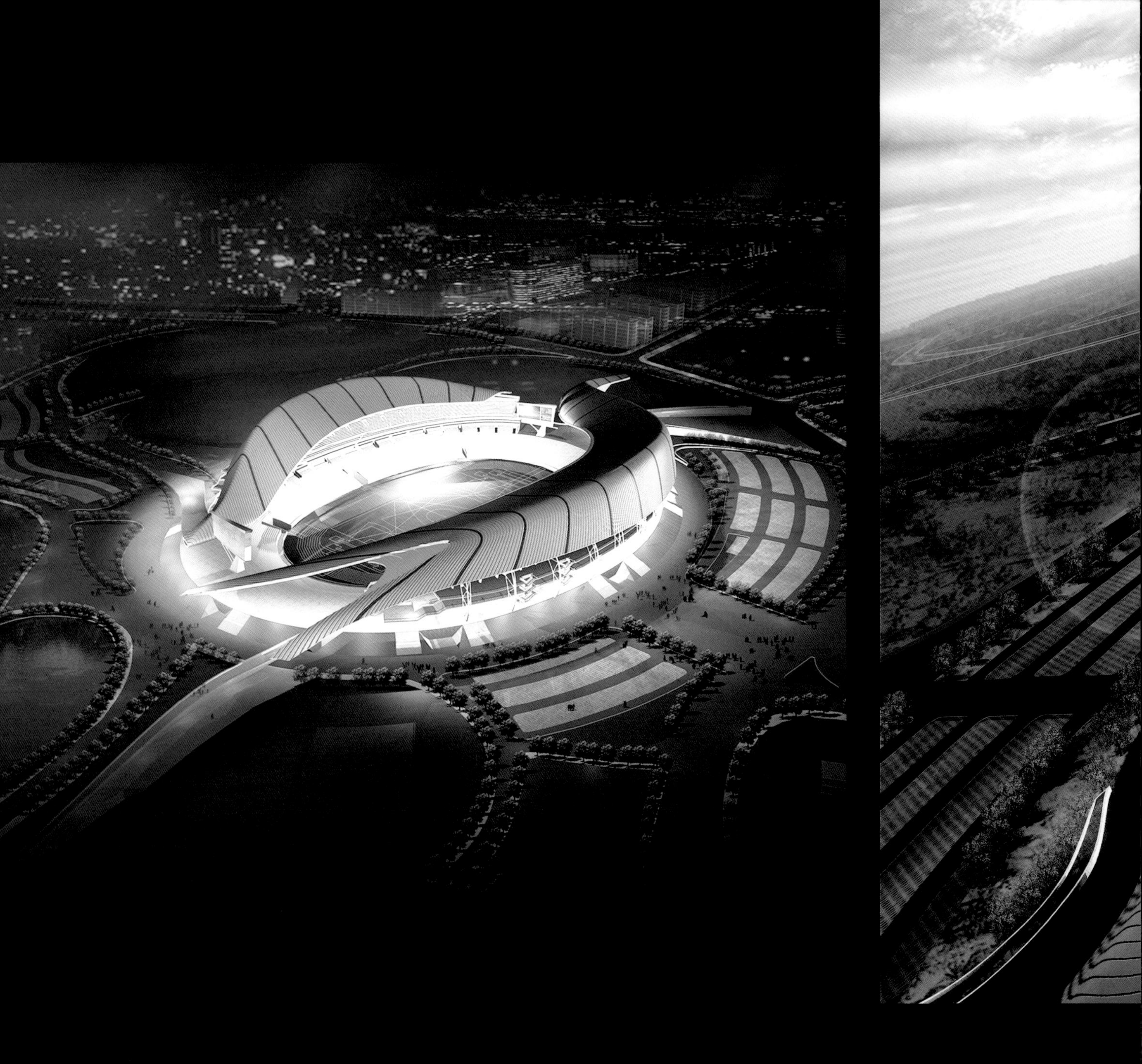

55 广州大学城体育馆
Gymnasium of Guangzhou Higher Education Mega Center (GZHEMC)

整个设计以保留基地内的自然山地景观为起始点，为了使建筑体量与周围连绵起伏的山峦相呼应，设计师参照山地的形成与地质的运动，构思了一个与自然规律十分贴近的“大地建筑”。在外观上暗示了一种向上的内力，这种内力为了向上突破，把地面撕开一道道的裂缝，建筑平缓地伏于地面，仿佛是延绵山体

The design starts from retaining the natural mountain landscape on the site. To echo the building volume with the surrounding undulating mountains, the designer conceives an "Earth Building" that is in perfect accordance with the natural laws, referencing the formation of mountain land and the geological movements. The appearance of the building suggests an upward-going internal force that breaks through and lacerates the ground. The building lying gently on the ground becomes practically a part of the

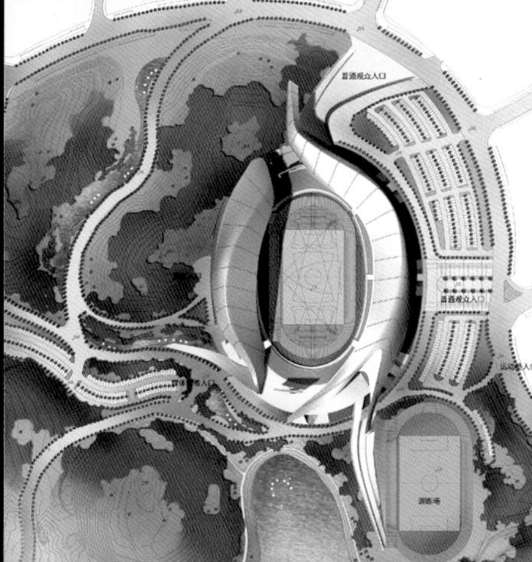
普通观众入口
普通观众入口

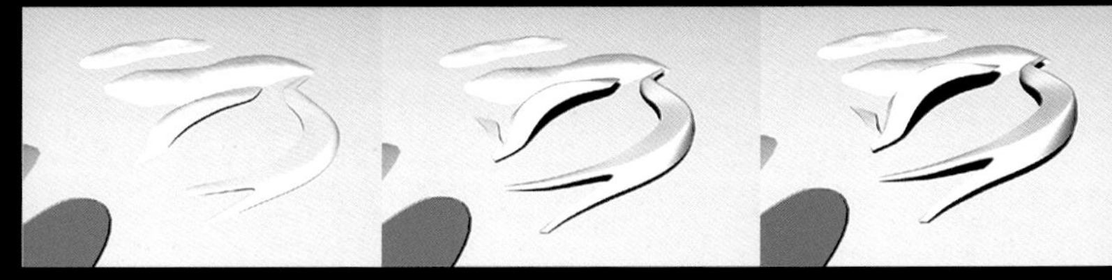

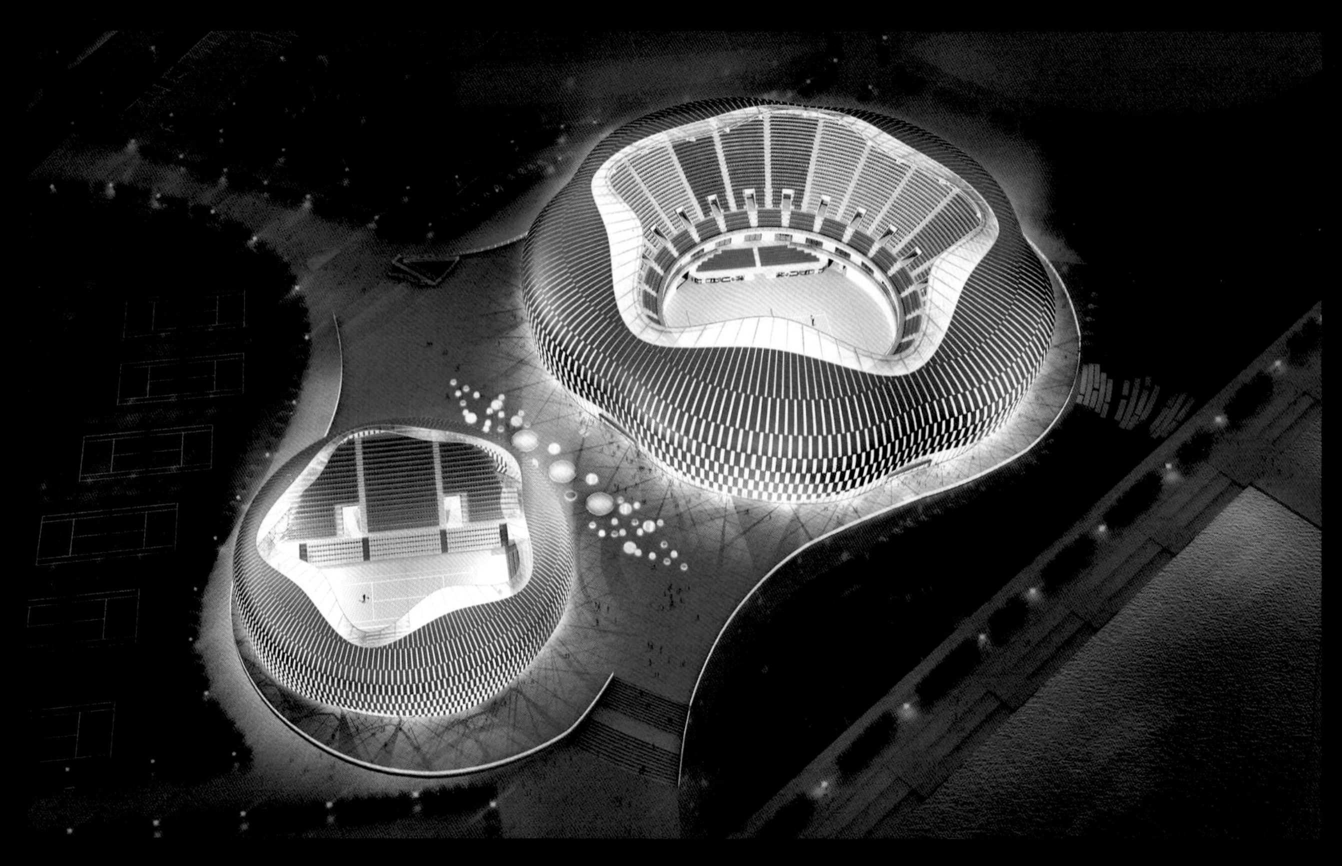

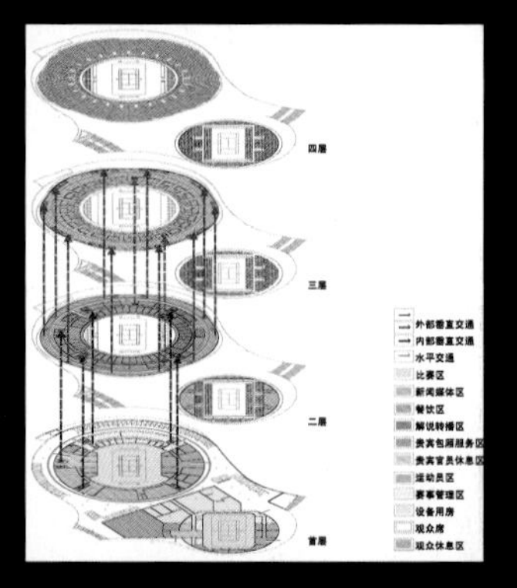

56 2010 年亚运会省属场馆网球中心
Tennis Center of Provincial Venues for Asian Games 2010, Guangzhou

项目主要由 10000 座的主比赛场，2000 座的副比赛场和 16 个室外网球训练场地组成，总建筑面积 33000 平方米，总投资约 1.65 亿元。

本设计以网球比赛冠军杯为设计概念，传达出“竞赛者对胜利的渴望”这一思维的表达，反映出网球中心的建筑个性。主网球场五瓣叶的造型犹如广州的市花木棉，通过融合、提炼，创造出外形大气、内在典雅的饱满形象。副场馆以同样的设计理念为基础，求同存异，使主副场馆遥相呼应、相得益彰。

本建筑的造型独特、简洁，空间形象具有鲜明的标识性，设计风格活泼、明快，立体造型通过不同材质的交融，表现出建筑的科技美和雕塑美。

The project consists of a 100,000-seat main venue, a 2,000-seat annexe and 16 outdoor tennis training fields, with a total floor area of 33,000 sqm and a total investment of about RMB 165 million yuan.

The design idea of the Tennis Center is derived from the tennis champion's cup. It conveys the competitors' *desire to win* and reflects the individuality of the building. The five-petal main venue resembles a kapok flower, the flower of Guangzhou city, presenting a mellow and full architectural image with grand appearance and elegant style. The annexe design implements the similar idea, so that the annexe could echo to and support the main venue in a consistent manner.

The Tennis Center is unique and concise in form, distinctive in space image, bright and sharp in design. The sculpture building volumes full of technological sense are created via the fusions of various materalities.

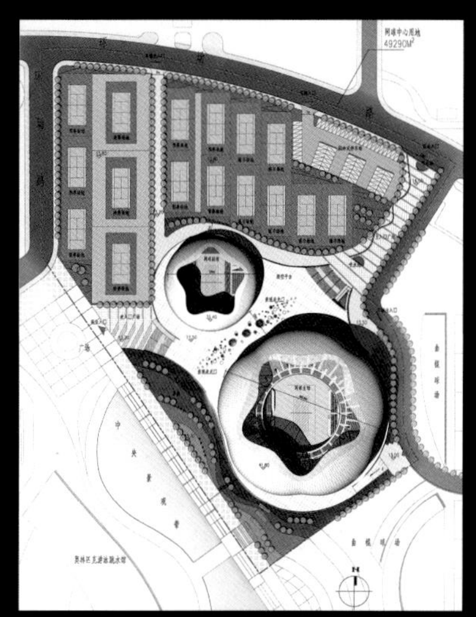
网球中心用地
49290M²

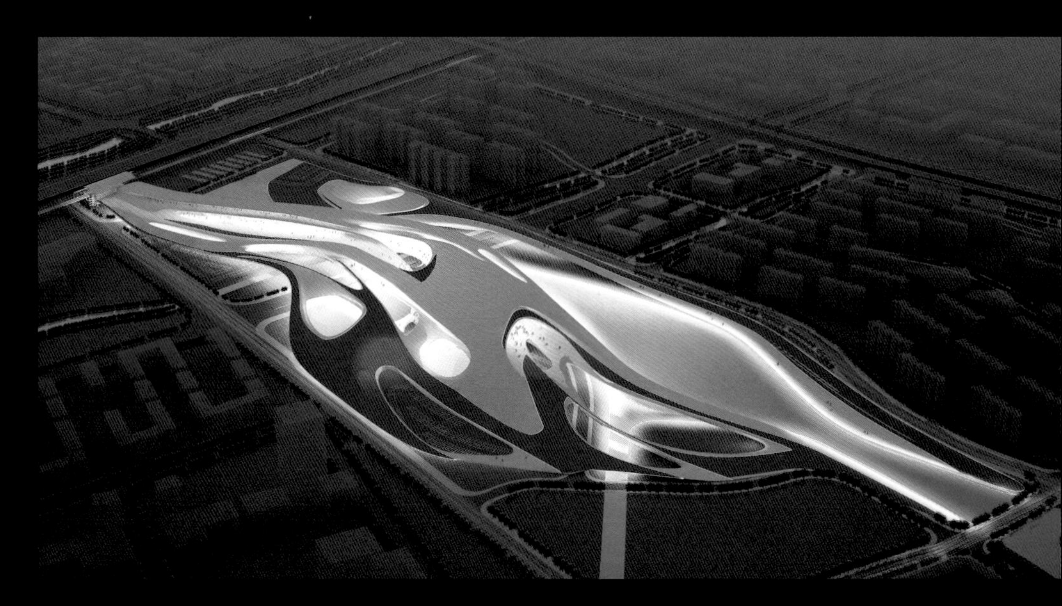

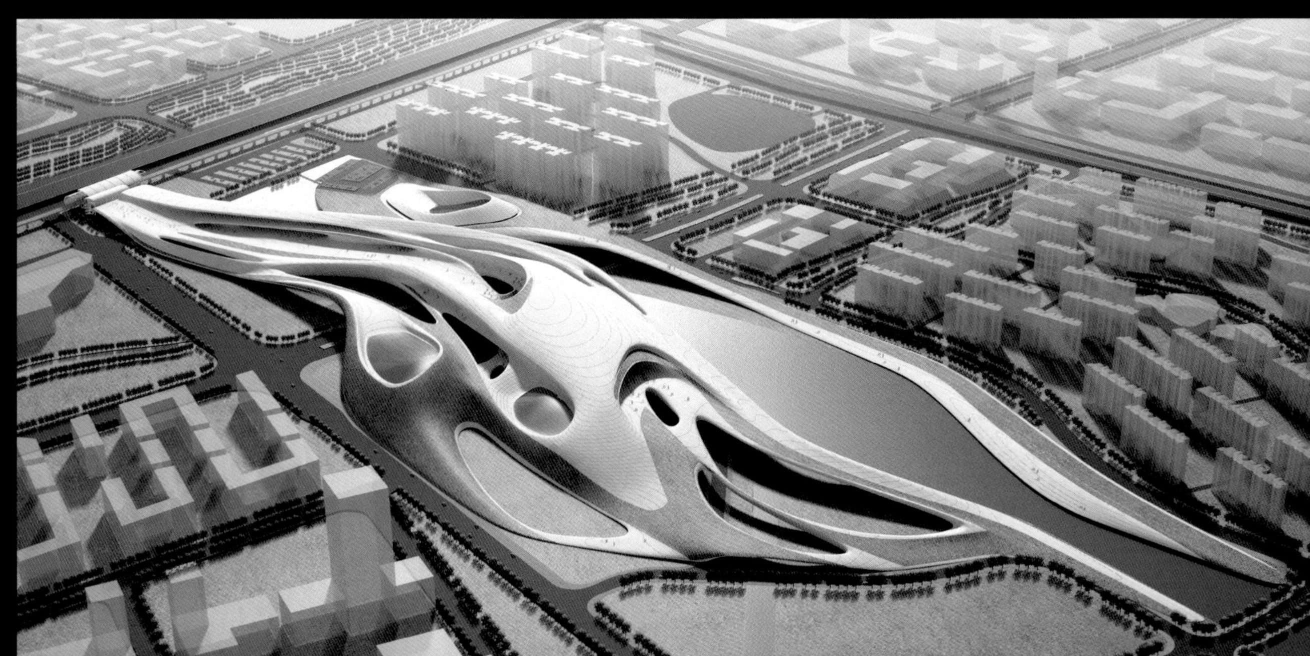

57 广州亚运体育综合馆区与媒体公共区（与扎哈·哈迪德建筑师事务所合作设计）

Guangzhou Asian Games Sports Complex and Media & Public Area

(In Collaboration with Zaha Hadid Architects)

本方案的设计灵感立足于通过构建一个“柔顺的，和谐的，充实的，绿色的”地标性建筑群，为广州新城增添一种令人愉悦的，充盈的氛围。

规划设计围绕着整个亚运城的流动优美的曲线，使得富有动感与生机的莲花山水道延伸到建筑设计本身，融为一体。观众，工作人员，运动员以及车辆在室外乃至室内的运动，与建筑的形态一道，如同流动的乐章，使得建筑与人和自然景观融为一个独特的有机体。

The design idea is to create a pleasant and vigorous atmosphere for Guangzhou New Town by creating a gentle, harmonious, substantiai and environment-friendly landmark building cluster.

The design and planning centers on the gracefully flowing curves of the Asian Games Town, making the dynamic and vital Lotus Mountain waterway extend to and integrate with the building itself. The indoor and outdoor movement of spectators, staff, athletes and vehicles, together with the architectural form composes a piece of moving music, producing a unique, cohesive whole that blends building, people and natural landscape into one.

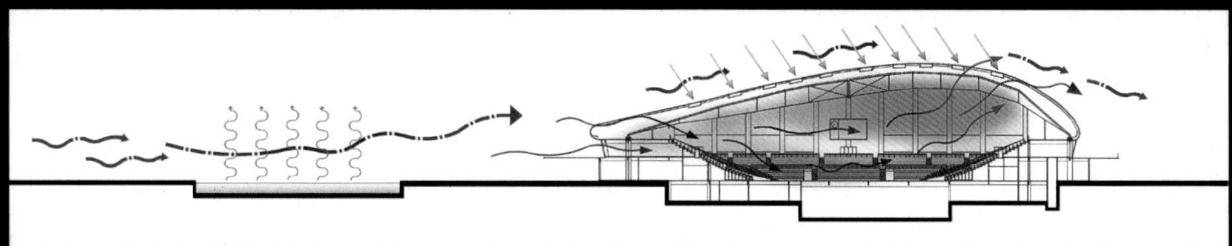

58 广州市萝岗区体育中心 Luogang Sports Center

项目总用地面积为 65430 平方米，其中包括一期体育中心项目用地 50430 平方米，二期广州国际羽毛球培训中心项目用地 15000 平方米。萝岗区体育中心项目的主要内容包括体育馆、游泳中心及体育场三部分。总建筑面积约为 2.2 万平方米，其中体育馆 20260 平方米（含地下室面积），观众总座位数为 5500 个（其中固定座位 3800 个）。游泳中心面积约为 1500 平方米，设露天观众席 1000 座。设 300 米跑道非标准体育场，面积约 300 平方米。

Luogang Sports Center occupies a total land area of 65,430 square meters, including 50,430 square meters for Sports Center (Phase Ⅰ) and 15,000 sqm for Guangzhou International Badminton Training Center (Phase Ⅱ). It mainly consists of gymnasium, swimming center and stadium, with a total floor area of 22,000 sqm, of which gymnasium accounts for 20,260 sqm (including basements), with a capacity of 5,500 seats in total (including 3,800 fixed seats). The swimming center occupies an area of about 1,500 sqm with 1,000 bleachers. In addition, a non-standard stadium with a 300-meter running track and an area of about 300 sqm is provided as well.

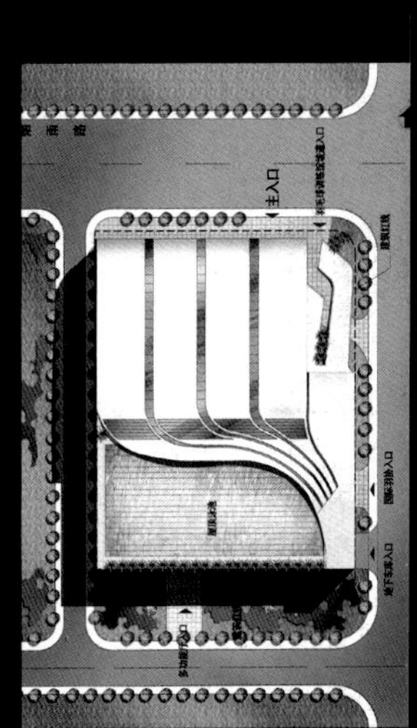

60 增城水上运动训练基地
Zengcheng Water Sports Training Base

建筑形象构思着力展现“水上运动”项目的特征——明快、轻盈、动感。

我们尊重基地的绿色植被，蓝色水面，本着绿色、环保、可持续发展的设计理念，我们最大限度地将其保留下来，作为建筑的背景。我们用蓝色的玻璃呼应水面，白色的实墙反衬绿色。蓝白对比的建筑简洁而又明快，大悬挑的造型隐喻着冲锋破浪的船首，又如昂首亮翅的白鹤。

The architectural image highlights the characteristics of water sports: bright, light, and dynamic.

Out of respect for the green vegetation and blue waters at site and in pursuit for the green, sustainable and environment-friendly philosophy, maximum efforts are made to preserve such vegetation and water areas as background of the building. Blue glass is employed to echo to the waters, and white walls to contrast with green color. The contrast between blue and white produces a clear-cut and bright architectural image. The large cantilevered element resembles a prow braving the wind and the waves, and also suggestive of a white crane with head high and wings wide.

综合技术经济指标一览表
1.运动员餐厅
2.运动员宿舍
3.综合教学办公大楼
4.综合训练馆及船库
5.游泳馆
6.学术交流接待中心
7.综合性球类馆
8.体育资源特色餐厅
9.广州海、陆、空模型培训中心
10.中国（南方）水上项目培训中心
11.码头船坞
总平面图

图书在版编目(CIP)数据

珠江设计33．上·汉英对照/广州珠江外资建筑设计院有限公司主编．—北京：中国建筑工业出版社，2014.1
ISBN 978-7-112-16294-9

Ⅰ.①珠… Ⅱ.①广… Ⅲ.①建筑设计-作品集-中国-现代 Ⅳ.①TU206

中国版本图书馆CIP数据核字（2014）第000399号

责任编辑：唐　旭　吴　绫
责任校对：张　颖　刘梦然
图书翻译：黄　捷　伍晖虹　韩　路　刘荣坤
翻　　译：梁　玲

珠江设计33（上）
广州珠江外资建筑设计院有限公司　主编
*
中国建筑工业出版社出版、发行（北京西郊百万庄）
各地新华书店、建筑书店经销
北京方舟正佳图文设计有限公司制版
北京顺诚彩色印刷有限公司印刷
*
开本：880×1230毫米　1/16　印张：12¾　字数：380千字
2014年2月第一版　2014年2月第一次印刷
定价：158.00元
ISBN 978-7-112-16294-9
(25036)